Pre-Milestone A and Early-Phase Systems Engineering

A Retrospective Review and Benefits for Future Air Force Systems Acquisition

Committee on Pre-Milestone A Systems Engineering: A Retrospective Review and Benefits for Future Air Force Systems Acquisition

Air Force Studies Board

Division on Engineering and Physical Sciences

NATIONAL RESEARCH COUNCIL
OF THE NATIONAL ACADEMIES

THE NATIONAL ACADEMIES PRESS
Washington, D.C.
www.nap.edu

THE NATIONAL ACADEMIES PRESS 500 Fifth Street, N.W. Washington, DC 20001

This is a report of work supported by Grant FA9550-06-1-0549 between the U.S. Air Force and the National Academy of Sciences. Any opinions, findings, conclusions, or recommendations expressed in this publication are those of the author(s) and do not necessarily reflect the view of the organizations or agencies that provided support for the project.

International Standard Book Number-13: 978-0-309-11475-2
International Standard Book Number-10: 0-309-11475-6

Limited copies of this report are available from:

Air Force Studies Board
National Research Council
500 Fifth Street, N.W.
Washington, DC 20001
(202) 334-3111

Additional copies are available from:

The National Academies Press
500 Fifth Street, N.W.
Lockbox 285
Washington, DC 20055
(800) 624-6242 or (202) 334-3313
(in the Washington metropolitan area)
Internet, http://www.nap.edu

Printed in the United States of America

THE NATIONAL ACADEMIES

Advisers to the Nation on Science, Engineering, and Medicine

The **National Academy of Sciences** is a private, nonprofit, self-perpetuating society of distinguished scholars engaged in scientific and engineering research, dedicated to the furtherance of science and technology and to their use for the general welfare. Upon the authority of the charter granted to it by the Congress in 1863, the Academy has a mandate that requires it to advise the federal government on scientific and technical matters. Dr. Ralph J. Cicerone is president of the National Academy of Sciences.

The **National Academy of Engineering** was established in 1964, under the charter of the National Academy of Sciences, as a parallel organization of outstanding engineers. It is autonomous in its administration and in the selection of its members, sharing with the National Academy of Sciences the responsibility for advising the federal government. The National Academy of Engineering also sponsors engineering programs aimed at meeting national needs, encourages education and research, and recognizes the superior achievements of engineers. Dr. Charles M. Vest is president of the National Academy of Engineering.

The **Institute of Medicine** was established in 1970 by the National Academy of Sciences to secure the services of eminent members of appropriate professions in the examination of policy matters pertaining to the health of the public. The Institute acts under the responsibility given to the National Academy of Sciences by its congressional charter to be an adviser to the federal government and, upon its own initiative, to identify issues of medical care, research, and education. Dr. Harvey V. Fineberg is president of the Institute of Medicine.

The **National Research Council** was organized by the National Academy of Sciences in 1916 to associate the broad community of science and technology with the Academy's purposes of furthering knowledge and advising the federal government. Functioning in accordance with general policies determined by the Academy, the Council has become the principal operating agency of both the National Academy of Sciences and the National Academy of Engineering in providing services to the government, the public, and the scientific and engineering communities. The Council is administered jointly by both Academies and the Institute of Medicine. Dr. Ralph J. Cicerone and Dr. Charles M. Vest are chair and vice chair, respectively, of the National Research Council.

www.national-academies.org

COMMITTEE ON PRE-MILESTONE A SYSTEMS ENGINEERING: A RETROSPECTIVE REVIEW AND BENEFITS FOR FUTURE AIR FORCE SYSTEMS ACQUISITION

PAUL G. KAMINSKI, Technovation, Inc., *Chair*
LESTER L. LYLES, U.S. Air Force (retired), *Vice Chair*
DEV A. BANERJEE, The Boeing Company
THOMAS W. BLAKELY, Lockheed Martin Aeronautics Company
NATALIE W. CRAWFORD, RAND Corporation
STEPHEN E. CROSS, Georgia Institute of Technology, Georgia Tech Research Institute
GILBERT F. DECKER, Independent Consultant
LLEWELLYN S. DOUGHERTY, Raytheon Company
JOHN V. FARR, Stevens Institute of Technology
JAMES H. FREY, Frey Associates
ROBERT A. FUHRMAN, Lockheed Martin Corporation (retired)
DAVID J. GORNEY, The Aerospace Corporation
JOHN M. GRIFFIN, Griffin Consulting
WESLEY L. HARRIS, Massachusetts Institute of Technology
RONALD T. KADISH, Booz Allen Hamilton
ROBERT H. LATIFF, Science Applications International Corporation
ALDEN V. MUNSON, JR., Office of the Director of National Intelligence
MARK K. WILSON, Mark Wilson Consulting

Staff

MICHAEL A. CLARKE, Board Director
JAMES C. GARCIA, Study Director
GREGORY EYRING, Senior Program Officer
CARTER W. FORD, Associate Program Officer
WILLIAM E. CAMPBELL, Senior Program Associate
LANITA R. JONES, Program Associate (until July 2007)
DETRA BODRICK-SHORTER, Administrative Coordinator

Preface and Acknowledgments

This study was requested by the Deputy Assistant Secretary of the Air Force for Science, Technology, and Engineering. The main goal was to examine the role that systems engineering can play during the defense acquisition life cycle in addressing the root causes of program failure, especially during the pre-Milestone A and early phases of a program. As chair and vice chair of the study committee, we extend special thanks to the committee members for their commitment and diligence, which enabled us to complete the task successfully.

This report has been reviewed in draft form by individuals chosen for their diverse perspectives and technical expertise, in accordance with procedures approved by the National Research Council's Report Review Committee. The purpose of this independent review is to provide candid and critical comments that will assist the institution in making its published report as sound as possible and to ensure that the report meets institutional standards for objectivity, evidence, and responsiveness to the study charge. The review comments and draft manuscript remain confidential to protect the integrity of the deliberative process. We wish to thank the following individuals for their review of this report:

Elliot I. Axelband, The RAND Corporation,
Dennis M. Buede, Innovative Decisions, Inc.,
Frank B. "Ted" Campbell, USAF (retired), Lockheed Martin Corporation,
Donald L. Cromer, USAF (retired), Hughes Space and Communications Company (retired),
Don R. Kozlowski, NAE, The Boeing Company (formerly McDonnell Douglas Corp.) (retired),
Annette J. Krygiel, Independent Consultant,

Louis S. Metzger, The MITRE Corporation,
William B. Rouse, NAE, Georgia Institute of Technology,
Andrew P. Sage, NAE, George Mason University,
John P. Stenbit, NAE, TRW, Inc. (retired),
A. Thomas Young, NAE, Lockheed Martin Corporation (retired), and
Gary Ziegler, Independent Consultant.

Although the reviewers listed above provided many constructive comments and suggestions, they were not asked to endorse the conclusions or recommendations, nor did they see the final draft of the report before its release. The review of this report was overseen by Thom J. Hodgson, NAE, North Carolina State University. Appointed by the National Research Council, he was responsible for making certain that an independent examination of this report was carried out in accordance with institutional procedures and that all review comments were carefully considered. Responsibility for the final content of this report rests entirely with the authoring committee and the institution.

The committee acknowledges and appreciates the contribution of the members of the Air Force Studies Board (AFSB) who developed the study statement of task in concert with the Air Force sponsor. The AFSB was established in 1996 as a unit of the National Research Council at the request of the United States Air Force. The AFSB brings to bear broad military, industrial, and academic scientific, engineering, and management expertise on Air Force technical challenges and other issues of importance to senior Air Force leaders. The board discusses potential studies of interest, develops and frames study tasks, ensures proper project planning, suggests potential committee members and reviewers for reports produced by fully independent ad hoc study committees, and convenes meetings to examine strategic issues. The board members, listed on page vi, were not asked to endorse the committee's conclusions or recommendations, nor did they review the final draft of this report before its release, although board members with appropriate expertise may be nominated to serve as formal members of study committees or as report reviewers.

The committee is very grateful to the Air Force sponsor for its dedicated support throughout the study and for the efforts of the National Research Council staff.

Paul G. Kaminski, *Chair*
Lester L. Lyles, *Vice Chair*
Committee on Pre-Milestone A Systems Engineering: A Retrospective Review and Benefits for Future Air Force Systems Acquisition

Contents

APPENDIXES

Acronyms

AEHF	advanced extremely high frequency
AETC	Air Education and Training Command
AFIT	Air Force Institute of Technology
AFMC	Air Force Materiel Command
AFMC/EN	AFMC Engineering Directorate
AFMCI	AFMC Instruction
AFSAB	Air Force Scientific Advisory Board
AFSC	Air Force Systems Command
AFSPC	Air Force Space Command
AIAA	American Institute of Aeronautics and Astronautics
AIT	Airborne Integrated Terminal
AMT	accelerated mission test
AoA	analysis of alternatives
APUC	average procurement unit cost
AUPP	average unit production price
C4ISR	command, control, communications, computers, intelligence, surveillance, and reconnaissance
CAD	computer-aided design
CAM	computer-aided manufacturing
CASE	computer-aided software engineering
CD	concept development
CDD	capability development document
CDR	critical design review
CIP	Component Improvement Program

CONOPS	concept of operations
COTS	commercial off-the-shelf
CPD	capabilities production document
CR	concept refinement
CRRA	capability review and risk assessment
CSE	Center for Systems Engineering
CSEP	certified systems engineering professional
DAB	Defense Acquisition Board
DAPA	Defense Acquisition Performance Assessment
DARPA	Defense Advanced Research Projects Agency
DAU	Defense Acquisition University
DCS	defense communications system
DE	domain experts
DNRO	Director of the National Reconnaissance Office
DOD	Department of Defense
DDAF	Department of Defense Architecture Framework
DODI	DOD Instruction
DOTMLPF	doctrine, organization, training, materiel, leadership, personnel, and facilities
DSB	Defense Science Board
DSCS	Defense Satellite Communications System
DSP	Defense Support Program
DT	development test
ENSIP	engine structural integrity program
ESC	Electronic Systems Center
EVM	earned value management
FAA	Federal Aviation Administration
FAB-T	Family of Advanced Beyond-Line-of-Sight Terminals
FCS	Future Combat Systems
FFP	firm fixed price
FFRDC	federally funded research and development center
FNA	functional needs analysis
FRPDR	full rate production decision review
FSA	functional solutions analysis
GAO	Government Accountability Office
GBS	Global Broadcast Service
GEO	geosynchronous Earth orbit
GMT	Ground Multiband Terminal

GOTS	government off-the-shelf
GPS	Global Positioning System
HC3	High Capacity Communications Capability
ICD	initial capabilities document
IDA	Institute for Defense Analyses
IDE	intermediate development education
IMS	Integration Master Schedule
INCOSE	International Council on Systems Engineering
IOC	initial operational capability
IPT	integrated product team
IRM	integrated risk management plan
JASSM	Joint Air-to-Surface Standoff Missile
JCIDS	Joint Capability Integration and Development System
JDAM	Joint Direct Attack Munition
JROC	Joint Requirements Oversight Council
JTRS	joint tactical radio system
KPP	key performance parameter
LCC	life cycle cost
LSI	lead systems integrator
M&S	modeling and simulation
MBMMR	Multi-Band Multi-Mode Radio
MCB	MJPO Configuration Board
MIG	MILSATCOM integration group
MILSATCOM	military satellite communications
MJPO	MILSATCOM Joint Program Office
MS	milestone
MUOS	mobile user objective system
NAE	National Academy of Engineering
NDIA	National Defense Industrial Association
NMS	national military strategy
NMT	Navy Multiband Terminal
NRO	National Reconnaissance Office
NSA	National Security Agency
NSS	national security strategy
NSSO	National Security Space Office

OEM	original equipment manufacturer
OJT	on-the-job training
OSD	Office of the Secretary of Defense
OSS&E	operational safety, suitability, and effectiveness
OT	operational test
OTA	Other Transactions Authority
OUSD (AT&L)	Office of the Under Secretary of Defense for Acquisition, Technology, and Logistics
PAUC	program acquisition unit cost
PBD	program budget decision
PDR	preliminary design review
PE	program element
PM	program manager
R&D	research and development
ROI	return on investment
S&E	scientist and engineer
S&T	science and technology
SATCOM	satellite communications
SBIRS	Space Based Infrared Systems
SDD	system design and development
SDOE	system design and operational effectiveness
SE	systems engineering
SEAG	systems engineering advisory group
SEI	Software Engineering Institute
SEIT	systems engineering integration team
SEM	systems engineering management
SEP	systems engineering plan
SES	Senior Executive Service
SETA	systems engineering and technical assistance
SMART-T	Secure Mobile Anti-Jam Reliable Tactical Terminal
SMC	Space and Missile Systems Center
SoS	system of systems
SOSCOE	System of Systems Common Operating Environment
SPO	system program office
SR	space radar
SRD	system requirements document
SSEA&I	System of Systems Engineering, Architecture and Integration
STE	staff technical equivalent
STSS	Space Tracking and Surveillance System
SWAMP	Software Acquisition Management Plan

T&E	test and evaluation
TD	technology development
TDS	technology development strategy
TMD	theater missile defense
TPM	technical performance measure
TRL	Technology Readiness Level
TSAT	Transformational Satellite Communications System
TSPR	total system performance responsibility
UFO	Ultra-High Frequency Follow-On
USAF	United States Air Force
USAFA	U.S. Air Force Academy
USD AT&L	Under Secretary of Defense for Acquisition, Technology and Logistics
WGS	Wideband Gapfiller Satellite
WIN-T	Warfighter Information Network-Tactical

Summary

Recent years have seen a serious erosion in the ability of U.S. forces to field new weapons systems quickly in response to changing threats, as well as a large increase in the cost of these weapons systems. Today the military's programs for developing weapons systems take two to three times longer to move from program initiation to system deployment than they did 30 years ago. This slowdown has occurred during a period in which threats have been changing more rapidly than ever and when technology advances and accumulated experience should have been accelerating rather than slowing the development process.

Many causes for this trend have been suggested, including the increased complexity of the tasks and the systems involved from both technological and human/organizational perspectives; funding instability; loss of "mission urgency" after the end of the Cold War; bureaucracy, which increases cost and schedule but not value; and the need to satisfy the demands of an increasingly diverse user community. The difficulty of focusing on a specific, homogeneous, post-Cold War threat made problems even worse. Yet although the suggested causal factors have merit, a common view is that better systems engineering (SE) could help shorten the time required for development, making it more like what it was 30 years ago.

Simply stated, SE is the translation of a user's needs into a definition of a system and its architecture through an iterative process that results in an effective system design. SE applies over the entire program life cycle, from concept development to final disposal. Figure S-1 illustrates the Department of Defense (DOD) acquisition life cycle.

The Committee on Pre-Milestone A Systems Engineering was tasked by the U.S. Air Force to examine the role that SE can play during the defense acquisi-

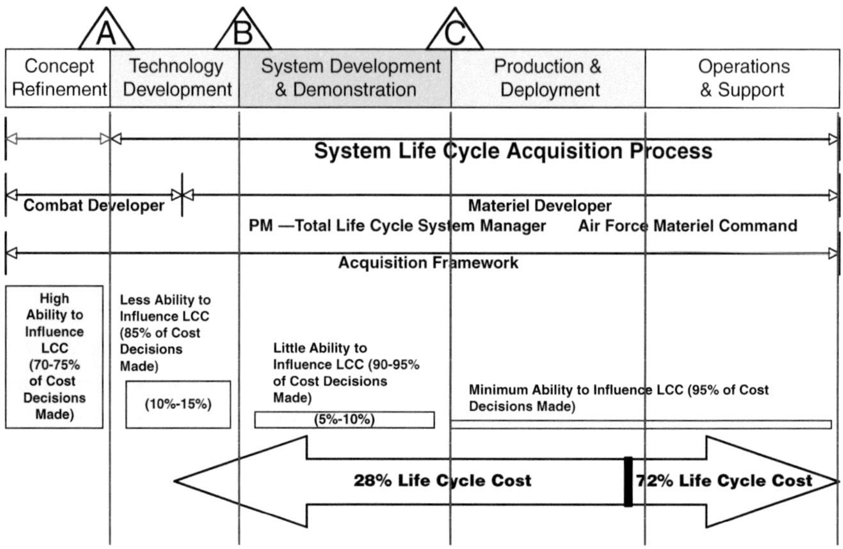

FIGURE S-1 DOD life cycle acquisition process. Points A, B, and C at the top of the figure represent Milestones A, B, and C. LCC, life cycle cost. SOURCE: Richard Andrews, 2003, *An Overview of Acquisition Logistics.* Fort Belvoir, Va.: Defense Acquisition University. Available at http://www.afcea.org/events/pastevents/documents/Track4Session4-AMCEmphasisonCustomerFocusedITInitiatives.ppt#364,12,Slide 12. Last accessed on November 20, 2007.

tion life cycle in addressing the root causes of program failure, especially during the pre-Milestone A and early phases of a program. Currently, few formal SE processes are applied to Air Force development programs before the Milestone A review.[1]

The committee has devoted much time and space in this report to trying to define a minimum set of systems engineering processes. Chapter 4, in particular, is devoted to this effort. The most important of these processes are summarized in the checklist in Box S-1 below in this summary (Box 4-1 in Chapter 4). A few of the things that need to be taken care of before Milestone A and just after it are the following: the consideration of alternative concepts (solutions) up front; the setting of clear, comprehensive key performance parameters (KPPs) and system requirements; and early attention to interfaces and interface complexity, to the concept of operations (CONOPS), and to the system verification approach. It is these early-stage processes that are covered in this report. The importance of

[1]This is a result of the elimination in the 1990s of the development planning function that had existed in the Air Force Systems Command.

stable requirements and funding between Milestone B and the achievement of initial operational capability (IOC) is stressed, as are processes including good configuration management and change control. The committee further stresses in the report what it regards as six of the most important process areas in its discussion of six "seeds of failure" in Chapter 4.

SYSTEMS ENGINEERING AND THE DOD ACQUISITION LIFE CYCLE

The use of formal systems engineering practices throughout the life cycle of an acquisition program is critical to fielding the required system on time and within budget. Across the top of Figure S-1 are the points at which important management decisions are made: Milestones A, B, and C. Concept development and refinement occur before Milestone A, and further technology development, to reduce system design and development (SDD) risk, occurs before Milestone B. Only after Milestone B does a program become an enterprise with dedicated funding. Importantly, Figure S-1 shows that about three-quarters of total system life cycle costs are influenced by decisions made before the end of the concept refinement phase at Milestone A, while about three-quarters of life cycle funds are not actually spent until after Milestone C. This means that although high-quality SE is necessary during the entire acquisition cycle, the application of SE to decisions made in the pre-Milestone A period is critical to avoiding (or at least minimizing) cost and schedule overruns later in a program. Much of the value of early, high-quality SE will be manifested as success in fulfilling Milestone B requirements.

MAIN FINDINGS AND RECOMMENDATIONS

The committee's main findings and recommendations are given below.

Finding. Attention to a few critical systems engineering processes and functions particularly during preparation for Milestones A and B is essential to ensuring that Air Force acquisition programs deliver products on time and on budget.

Today's weapons systems provide unprecedented capabilities but also involve complex interfaces with external command, control, and communications systems and rely on a greater volume of software than ever before. Early decisions on the weapons system requirements and capabilities have a disproportionately large impact on program cost and schedule. The committee also recognizes that a lack of flexibility (a result of overly rigid processes or a lack of trust among program participants or stakeholders) can limit the ability of a program manager to change early decisions that warrant changing.

The committee found many gaps and inconsistencies in the way that the Air Force manages pre-Milestone A activities. The committee heard from presenters of some cases for which required documents were completed pro forma and filed away, never to be seen again, or for which required steps were skipped completely. The current practice of initiating programs at Milestone B denies the acquisition review authority the earlier opportunity (at Milestone A) to make judgments about the maturity of the technologies on which the program is based and to decide whether technologies need to be further developed prior to making a Milestone B commitment to system development and demonstration.

Recommendation. The Air Force leadership should require that Milestones A and B be treated as critical milestones in every acquisition program and that a checklist such as the "Pre-Milestone A/B Checklist" suggested by the committee (see Box S-1 in this Summary) be used to judge successful completion.

A rigorous, standard checklist of systems engineering issues should be addressed by each program through both the pre-Milestone A and pre-Milestone B phases. The committee's recommended 20-item checklist is shown in Box S-1.

While the committee considers that each item on the checklist is important, it calls attention to several items that warrant further discussion. Item 2 recognizes that the world changes too fast to be friendly to long development cycles. The committee believes that the Air Force should strive to structure major development programs so that initial deployment is achieved within, say, 3 to 7 years. Thirty years ago, this was a typical accomplishment—for example, nearly 40 years ago, the Apollo program put the first man on the Moon in fewer than 8 years.

The development time issue is addressable by applying systems engineering to Items 3, 4, and 13 through 15 before Milestones A and B. The definition of clear KPPs by Milestone A and clear requirements by Milestone B that can remain stable through IOC can be essential to an efficient development phase. It is also important that critical technologies be sufficiently mature prior to starting SDD. The committee observed that although today's systems are not necessarily more complex internally than those of 30 years ago, their "external complexity" often is greater, because today's systems are more likely to try to meet many diverse and sometimes contradictory requirements from multiple users. This kind of complexity can often lead to requirements being changed between Milestone B and IOC, and it can lead to relying on immature technology.

Item 19 of the checklist stresses the importance of placing experienced, domain-knowledgeable managers in key program positions. The committee has observed that many of the truly extraordinary development programs of the past, such as Apollo, the Manhattan Project, the early imaging satellite programs, the U-2, the fleet ballistic missile system, and nuclear submarines, were managed by relatively small (and often immature) agencies with few established processes and controls. In that environment, dedicated managers driven by urgent missions accomplished feats that often seem incredible today.

BOX S-1
Pre-Milestone A/B Checklist

Concept Development

1. Have at least two alternative concepts to meet the need been evaluated?

   The purpose of alternatives is to stimulate thinking to find the simplest, fastest, and cheapest solution.

2. Can an initial capability be achieved within the time that the key program leaders are expected to remain engaged in their current jobs (normally less than 5 years or so after Milestone B)? If this is not possible for a complex major development program, can critical subsystems, or at least a key subset of them, be demonstrated within that time frame?

   Achieving capabilities or demonstrating critical subsystems while key program leaders remain engaged is important to get the capability into service quickly and cost-effectively and to begin the process of incremental improvements based on operational experience.

3. Will risky new technology have been matured before Milestone B? If not, is there an adequate risk mitigation plan?

   The development of risky new technology in parallel with a major development program can be costly in terms of both time and money.

4. Have external interface complexities (including dependencies on other programs) been identified and minimized? Is there a plan to mitigate their risks?

   Complex, ill-defined, external requirements and interfaces can be a major source of requirements instability during the development phase. This can be of particular importance when a system must operate in a system-of-systems environment.

Key Performance Parameters and CONOPS

5. At Milestone A, have the KPPs been identified in clear, comprehensive, concise terms that are understandable to the users of the system?

   It is important that KPPs be expressed in terms understandable to all of the stakeholders. Failure to define the system's KPPs simply and clearly at Milestone A is a first step to requirements instability and overruns later.

continued

BOX S-1 Continued

6. At Milestone B, are the major system-level requirements (including all KPPs) defined sufficiently to provide a stable basis for the development through IOC?

   Beginning development without a complete list of stable requirements is one of the key "seeds of failure" described in Chapter 4 in this report. It is important to complete requirements trade-offs prior to the development phase.

7. Has a CONOPS been developed showing that the system can be operated to handle the expected throughput and meet response time requirements?

   It can be costly to discover too late that the system as designed cannot be operated to meet its requirements.

Cost and Schedule Scoping

8. Are the major known cost and schedule drivers and risks explicitly identified, and is there a plan to track and reduce uncertainty?

   Identifying the major cost and schedule risk areas, with particular attention to this checklist and the six seeds of failure—inexperienced leadership, external interface complexity, system complexity, incomplete requirements at Milestone B, immature technology, and high reliance on new software—can help focus management on these issues early.

9. Has the cost confidence level been accepted by the stakeholders for the program?

   It is important that stakeholders understand the degree of risk so that the stakeholders will not disrupt the program as inevitable development program surprises unfold later on. It will generally not be possible by Milestone A or Milestone B to identify all the risk areas that might surface later in a development program, but a frank, early disclosure of known potentials for risk can help sustain stakeholder support later on.

Performance Assessment

10. Is there a sufficient collection of models and an appropriate simulation environment to validate the selected concept and the CONOPS against the KPPs?

continued

BOX S-1 Continued

In large, complex programs, the development of models early on can be very important to later management of requirements changes and performance verification.

11. At Milestone B, do the requirements take into account likely future mission growth over the program life cycle?

    The committee advocates freezing new requirements and new technology insertion after Milestone B but also notes that making provisions in the initial requirements to facilitate later upgrades could have great long-term value.

Architecture Development

12. Has the system been partitioned to define segments that can be independently developed and tested to the greatest degree possible?

    Effective partitioning of a complex system can greatly reduce its development cost.

13. By Milestone A, is there a plan to have information exchange protocols established for the whole system and its segments by Milestone B?

    Such a plan developed early on can greatly reduce interface problems later in the development phase when they would be more difficult and costly to fix.

14. At Milestone B, has the government structured the program plan to ensure that the contractor addresses the decomposition of requirements to hardware and software elements sufficiently early in the development program?

    The histories of programs with cost and schedule overruns are replete with examples of large software developments that had to be redone because requirements from the hardware side were assigned or determined late.

Risk Assessment

15. Have the key risk drivers (not only the technology drivers) been identified?

    Identifying and managing risk early can pay large dividends; it is important to focus on the six "seeds of failure" (see item 8 above).

continued

BOX S-1 Continued

Program Implementation Strategy

16. Does the government have access over the life of the program to the talent required to manage the program? Does it have a strategy over the life of the program for using the best people available in the government, the FFRDCs, and the professional service industry?

    Seasoned management is critical; the government's job is to find the best!

17. At Milestone A, is there a plan defining how the pre-Milestone B activity will be done, and by whom?

    Identifying the program and system managers early, identifying the FFRDC or SETA support needed, thinking through the use of competitive system concept contracts—all can have a decisive impact on the government's ability to select the best concept, to define by Milestone B system requirements that can remain stable through IOC, and to select the best development contractors.

18. Is there a top-level plan for how the total system will be integrated and tested?

    A well-thought-out strategy for verifying system performance, including optimum phasing of verification tests throughout the assembly process, and well-thought-out use of analytical models and external simulators can have a large positive impact on ultimate cost, schedule, and performance.

19. At Milestone B, have sufficiently talented and experienced program and systems engineering managers been identified? Have they been empowered to tailor processes and to enforce requirements stability from Milestone B through IOC?

    Seasoned leaders in these areas are critical to maintaining focus and discipline through IOC.

20. Has the government attempted to align the duration of the program manager's assignment with key deliverables and milestones in the program?

    A combination of assignment extension and time-certain milestones will help align incentives.

NOTE: KPP, key performance parameter; CONOPS, concept of operations; IOC, initial operational capability; FFRDC, federally funded research and development center; SETA, systems engineering and technical assistance.

The committee believes that the accumulation of processes and controls over the years—well meant, of course—has stifled domain-based judgment that is necessary for timely success. Formal SE processes should be tailored to the application. But they cannot replace domain expertise. In connection with item 19, the committee recommends that the Air Force place great emphasis on putting seasoned, domain-knowledgeable personnel in key positions—particularly the program manager, the chief system engineer, and the person in charge of "requirements"—and then empower them to tailor standardized processes and procedures as they feel is necessary.

One key pre-Milestone A task is the analysis of alternatives (AoA), which entails evaluating alternative concepts and comparing them in terms of capabilities, costs, risks, and so on. Checklist items 1 through 4, 12, and 13 should be completed before the AoA, while items 5 through 11 and 14 through 20 may be addressed after the AoA.

Finding. The creation of a robust systems engineering process is critically dependent on having experienced systems engineers with adequate knowledge of the domain relevant to a contemplated program.

While the systems engineering process is, broadly, reusable, it depends on having domain experts who are aware of what has gone wrong (and right) in the past recognize the potential to repeat the successes under new circumstances and avoid repeating the errors.

Ideally, a person or persons with domain knowledge would have had experience working on exactly the same problem, or at least a problem related to the one at hand. If that is not so (and it might not be if the problem has never been addressed before, as was the case for Apollo and nuclear submarines), the term could be taken to refer to academic training in the relevant field of engineering or science. It would also refer to the practice in critical thinking and problem solving that comes with learning to be a systems engineer and then building on that foundation to gain the experiential knowledge and understanding of engineering in the context of an entire system. Systems engineering is enabled by tools that have been developed to assist in the management of systems engineering (not to be confused with the *practice* of systems engineering).

Both industry and Air Force presenters told the committee that there are not enough domain-knowledgeable and experienced systems engineers to support all of the programs that need them.

Recommendation. The Air Force should assess its needs for officers and civilians in the systems engineering field and evaluate whether either its internal training programs, which include assignments on Air Force programs that provide mentoring by experienced people and hands-on experience in the application of systems engineering principles, or external organizations are able to produce the required quality and quantity of systems engineers and systems engineering skills. Based on this assessment, the Air Force first should determine how and where students should be trained, in what numbers, and at what cost, and then implement a program that meets its needs.

The Air Force needs to attract, develop, reward, and retain systems engineers across the full spectrum of relevant domains, engage them in the early (pre-Milestone A) phase of new programs (or modification programs), and sustain their participation throughout the life of the programs. One important step in this process would be to create an Air Force occupational code for systems engineering so that engineers' experience and education can be tracked and managed more effectively. The Air Force should support an internal systems engineering career track that rewards the mentoring of junior systems engineering personnel, provides engineers with broad systems engineering experience, provides appropriate financial compensation to senior systems engineers, and enables an engineering career path into program management and operations.

Finding. The government, federally funded research and development centers (FFRDCs), and industry all have important roles to play throughout the acquisition life cycle of modern weapons systems.

Since the need for a new or upgraded weapons system is most often first recognized by the military user, it is appropriate for the military to codify its requirements and, with support from FFRDCs and independent systems engineering and technical assistance (SETA) contractors, to explore materiel and nonmateriel solutions (such as doctrinal, organizational, or procedural changes) as well as to assess the potential for new technology to provide enhanced capabilities. While it is appropriate and usually desirable to engage development contractors in the pre-Milestone B process using competitive study contracts, the source selection for system development and demonstration should not be made until after the work associated with Milestones A and B is complete.

Recommendation. Decisions made prior to Milestone A should be supported by a rigorous systems analysis and systems engineering process involving teams of users, acquirers, and industry representatives.

Working together, government and industry can develop and explore solutions using systems engineering methodology to arrive at an optimal systems solution.

Finding. The Air Force used to have a development planning organization that applied pre-Milestone A systems engineering processes to a number of successful programs, but that organization was allowed to lapse.

The role of the Air Force development planning organization, which was within the Air Force Systems Command, was to provide standard evaluation tools and perform pre-Milestone A systems engineering functions across acquisition programs. The early 1990s saw an erosion of this front-end planning organization along with its funding as the Air Force Systems Command (now the Air Force Materiel Command) began to play a decreasing role in program execution. In the opinion of several speakers who met with the committee, one main reason for the erosion of funding was a lack of congressional support for the planning function.

Recommendation. A development planning function should be established in the military departments to coordinate the concept development and refinement phase of all acquisition programs to ensure that the capabilities required by the country as a whole are considered and that unifying strategies such as network-centric operations and interoperability are addressed.

The Air Force and the other military services should establish a development planning organization like that which existed in the early 1990s.

The roles and functions of the various organizations involved in acquiring major weapons systems need to be clearly defined. The responsibility for executing systems engineering and program management in the pre-Milestone A and B phases should be vested in the military departments that do the actual development planning functions. This should not be the responsibility of the Office of the Secretary of Defense (OSD) or of the Joint Staff. Instead, those offices need to enable the creation and functioning of military department development planning organizations with policy measures and, where appropriate, resources. The Joint Staff, under the auspices of the Joint Requirements Oversight Council (JROC), may help to define the requirements for major programs in the course of the development planning process, but it should not run the process itself.

The existence of "joint" programs or a program such as Missile Defense, which has several related systems being developed by different military services, requires clear guidance from both OSD and the Joint Staff about who is in charge. These programs need to be harmonized and integrated by the responsible integrating agency. However, development planning activities should still take place in the military departments where the expertise resides. Consequently, the development planning should be managed by that agency.

While this committee cannot predict how Congress will view the revival of a good planning process to support pre-Milestone A program efforts, it is still important for the Air Force and DOD to make the case for the critical importance of this process before Congress and others. A development planning process is important not to start new programs, but rather to ensure that any new program (or a new start of any kind) is initiated with the foundation needed for success. Funding for this planning function needs to be determined by the military services, including both the acquisition communities and those (the warfighters) who generate the operational requirements.

CONCLUDING THOUGHTS

Many of the conclusions reached and recommendations made by the committee are similar to those of previous reviews. Most of the past recommendations were never implemented, so one of this committee's most critical thoughts relates to the importance of implementation. A sampling of key findings and recommendations from previous studies follows:

- Government Accountability Office (GAO) [2,3]
  - Separate technology development from systems acquisition. Commit to a program only if the technology is sufficiently mature. Set the minimum Technology Readiness Level (TRL).
  - Stabilize the requirements early.
  - Employ systems engineering techniques before committing to product development.
  - Employ evolutionary approaches that pursue incremental increases in capability.
  - Address shortfalls in science, engineering, and program management staff.
- National Defense Industrial Association (NDIA)[4]
  - Increase SE awareness and recognize SE authority in the program formulation and decision process.
  - Incentivize career SE positions within the government.
- Defense Science Board (DSB)[5]
  - Overhaul the requirements process.
  - Stabilize acquisition tours.
  - Establish a robust SE capability.
- Defense Acquisition Performance Assessment (DAPA)[6]
  - Strategic technology exploitation is a key U.S. advantage. Opportunities need to be identified early.
  - The U.S. economic and security environments have changed—for example, there are fewer prime contractors, smaller production runs, reduced plant capacity, fewer programs, and unpredictable threats.
  - The acquisition system must deal with instability of external funding.

---

[2]Government Accountability Office (GAO), 2003, *Defense Acquisitions: Improvements Needed in Space Systems Acquisition Management Policy*, September. Available at http://www.gao.gov/new.items/d031073.pdf. Last accessed April 2, 2007.

[3]GAO, 2005, *Space Acquisitions: Stronger Development Practices and Investment Planning Needed to Address Continuing Problems*, July. Available at http://www.gao.gov/new.items/d05891t.pdf. Last accessed April 2, 2007.

[4]National Defense Industrial Association (NDIA) Systems Engineering Division, 2003, *Task Report: Top Five Systems Engineering Issues in Defense Industry*, January, Arlington, Va.: NDIA.

[5]Defense Science Board/Air Force Scientific Advisory Board Joint Task Force, 2003, *Acquisition of National Security Space Programs*, May, Washington, D.C.: OUSD (AT&L).

[6]Ronald Kadish, Gerald Abbott, Frank Cappuccio, Richard Hawley, Paul Kern, and Donald Kozlowski, 2006, *Defense Acquisition Performance Assessment*. Available at http://www.acq.osd.mil/dapaproject/documents/DAPA-Report-web/DAPA-Report-web-feb21.pdf. Last accessed on April 2, 2007.

—The DOD management model is based on a lack of trust. Quantity of review has replaced quality. There is no clear line of responsibility, authority, or accountability.
—Oversight is preferred to accountability.
—Oversight is complex, not process- or program-focused (as it should be).
—The complexity of the acquisition process increases costs and draws out the schedule.
—Incremental improvement applied solely to the "little a" acquisition process[7] requires all processes to be stable—but they are not.

The committee notes that successful implementation of these recommendations requires the "zipper concept"—making connections at all levels, from the senior leadership of the Air Force and DOD down to the working levels within key program management offices and supervisory staffs.

[7]The Acquisition—"Big A"—system is often believed to be a simple construct that efficiently integrates three independent processes: requirements, budgeting, and acquisition. "Little a," on the other hand, refers to the acquisition process that focuses on "how to buy" in an effort to balance cost, schedule, and performance; it does not include requirements and budgeting.

1

Introduction and Overview

The time required to execute large, government-sponsored systems development programs has more than doubled over the past 30 years, and the cost growth has been at least as great. Many causes for this trend have been suggested, including the increased complexity of the systems involved; instability of external funding; loss of "mission urgency" after the end of the Cold War; diminished depth of talent in the government and contractor community; requirements creep; the need to satisfy the demands of an increasingly diverse user community; inadequate up-front project planning; lack of management oversight, accountability, and clear metrics on both the government and contractor sides; and the exponential growth in, and reliance on, software. Nevertheless, this trend is particularly puzzling given the enormous productivity advantages conferred by the advent of the Internet and e-mail, the revolution in electronics and computer technology, and advances in knowledge-management and collaboration tools such as computer-aided design (CAD), computer-aided manufacturing (CAM), computer-aided software engineering (CASE), and modeling and simulation.

During World War II and the early Cold War, programs such as the Manhattan Project, the Defense Support Program (DSP), the intercontinental ballistic missile, and the U-2 surveillance aircraft all delivered very quickly, generally in fewer than 6 years, first products that today would be described as major systems. Currently, such major programs would likely require 10 to 20 years to complete. System complexity has grown dramatically, and products are not delivered under the same technological, human, and organizational guidelines as before. There has been about a threefold increase in delivery time for most major systems. Table 1-1 summarizes some well-known examples comparing historical program accomplishments with those of more recent programs. Table 1-2 shows some

TABLE 1-1 Representative Development Times of Major Historical and Recent Programs

Program and Year of First Use	Years to First Use from Contractor Selection
Historical programs	
Manhattan Project (1945)	2½
Defense Support Program (1970)	5½
Intercontinental Ballistic Missile (1958)	3½
Apollo (1967)	8
F-104 (1958)	5
SR-71 (1962)	3
Recent programs	
Future Imagery Architecture-Optical	13 (projected when canceled)
Space Based Infrared Systems/Boost Surveillance and Tracking System (to be determined)	>20
B-2 bomber (1993)	11
Joint Strike Fighter (to be determined)	≈13
F-22 (2005)	14

TABLE 1-2 Cost and Schedule Outcomes Sorted by Percentage of Product Development Remaining

Program	Cost Growth (Percent)[a]	Schedule Growth (Months)	Development Remaining (Percent)
Aerial Common Sensor	45	24	85
Future Combat System	48	48	78
Joint Strike Fighter	30	23	60
Expeditionary Fighting Vehicle	61	48	49
C-130 Avionics Modernization	122	Delays anticipated	Undetermined
Global Hawk (RQ-4B)	166	Delays anticipated	Undetermined

[a]Cost growth is expressed as the percentage change in program development cost estimates in 2005 base-year dollars.

SOURCE: Reprinted from Government Accountability Office, 2006, *Major Weapon Systems Continue to Experience Cost and Schedule Problems Under DOD's Revised Policy*, GAO-06-368, April. Available at http://www.gao.gov/highlights/d06368high.pdf. Last accessed April 2, 2007.

modern programs to further illustrate the current trend toward increasing cost and time to deployment.

In an effort to develop consistent policies and methodologies to address cost and schedule overruns, the Department of Defense (DOD) has published numerous policies, undertaken studies,[1] and developed several guidebooks such as the

[1]See Defense Science Board, 2007, *21st Century Strategic Technology Vectors, Vol. I: Accelerating the Transition of Technologies into US Capabilities*, April, Washington, D.C.: OUSD (AT&L).

5000 series, a Systems of Systems Guidebook, and the Systems Engineering Plan Guidebook.[2]

The individual services and intelligence agencies have also published policies and guides to supplement the DOD policies and to develop service- and agency-specific processes. For example, the National Reconnaissance Office (NRO) has made attempts over the past few years to address some of the acquisition difficulties that it has experienced. It has emphasized a more rigorous adherence to milestone decision gates and has made the extensive use of independent reviews of program readiness a necessary step before proceeding to the next phase of a program. It has also modified its acquisition schedules to align major decisions more closely with the results of major design reviews, and mandated more frequent post-Milestone C reviews by the decision authority. On a more technical level, the NRO, in cooperation with its industry team members, has reinstituted a minimum essential set of specifications and standards on such diverse topics as systems engineering (SE) and the qualification of key components.

Yet despite the myriad of new and revised processes throughout government acquisition organizations, there is little sign that performance is returning to the development productivity that was achieved decades ago. Indeed, one is tempted to conclude that performance diminishes as procurement organizations mature and their processes become more complex. This is counter to the trend in the private sector, where automobiles, commercial aircraft, commercial spacecraft, and consumer electronics have experienced 50 to 70 percent reductions in cycle times.[3]

Recent studies done by the Government Accountability Office (GAO) have expressed continuing concern about program cost and schedule growth problems, even under the revised policies being promulgated by the DOD. As the GAO stated in 2006:

> Changes made in DoD's acquisition policy over the past 5 years have not eliminated cost and schedule problems for major weapons development programs. Of the 23 major programs we assessed, 10 are already expecting development cost overruns greater than 30 percent or have delayed the delivery of initial operational capability to the warfighter by at least 1 year. The overall impact of these costly conditions is a reduction in the value of DoD's defense dollars and a lower return on investment. Poor execution of the revised acquisition policy is a major cause of DoD's continued problems. The DoD frequently bypasses key steps of the knowledge-based process outlined in the policy, falls short of attaining key knowledge, and continues to pursue revolutionary—rather than evolutionary or incremental—advances in capability. Nearly 80 percent of the programs GAO reviewed did not fully follow the knowledge-based process to develop a sound

---

[2] Defense Acquisition University, 2007, *Systems and Software Engineering Publications and Documents*. Available from http://www.acq.osd.mil/se/publications.htm. Last accessed on May 2, 2007.

[3] See Defense Science Board, 2007, *21st Century Strategic Technology Vectors, Vol. I: Accelerating the Transition of Technologies into US Capabilities*, April, Washington, D.C.: OUSD (AT&L).

> business case before committing to systems development. Most of the programs we reviewed started system development with immature technologies, and half of the programs that have held design reviews did so before achieving a high level of design maturity. These practices increase the likelihood that problems will be discovered late in development when they are more costly to address. Furthermore, DoD's continued pursuit of revolutionary leaps in capability also runs counter to the policy's guidance. The DoD has not closed all of the gaps in the policy that GAO identified nearly 3 years ago, particularly with regard to adding controls and criteria. Effective controls require decision makers to measure progress against specific criteria and ensure that managers capture key knowledge before moving to the next acquisition phase. However, DoD's policy continues to allow managers to approach major investment decisions with many unknowns. Without effective controls that require program officials to satisfy specific criteria, it is difficult to hold decision makers or program managers accountable to cost and schedule targets. In this environment, decision-making transparency is crucial, but DoD is lacking in this area as well.[4]

The Air Force and DOD are concerned about the impact that this trend is having in terms of the cost of fielding new systems, the erosion of spending power, and perhaps more importantly, the loss of agility to respond to rapidly changing threats.

As suggested above, programs may fail or exhibit cost and schedule overruns for many reasons. Some of these are external to the program, such as funding instability; others are internal to the program and thus under the control of DOD managers. Two critical factors in the success or failure of programs that fall in the latter category are the need for high-quality systems engineering and the related issue of the need for a high-quality systems engineering workforce. These success factors are the focus of this report and are described further below.

SYSTEMS ENGINEERING

The complexity of systems that humans choose to build, manage, and control continues to grow, with no sign of letting up—outpacing the ability of engineers to develop processes and tools to manage their development. The committee considered the degree to which growth in complexity is responsible for the before-mentioned increases in the cost and time required to deploy new systems. It found a number of legacy programs that appear to be as complex as or more complex than follow-on programs developed more recently at much greater cost and time. The early imaging satellites and the DSP are examples of such early successful programs, while the Space Based Infrared Systems (SBIRS) program (DSP follow-on) has experienced numerous cost and schedule overruns (see

[4]Government Accountability Office, 2006, *Major Weapon Systems Continue to Experience Cost and Schedule Problems Under DOD's Revised Policy*, GAO-06-368, April. Available at http://www.gao.gov/highlights/d06368high.pdf. Last accessed on April 2, 2007.

Chapter 2). The Apollo program in the 1960s, arguably one of the most complex space programs ever, took fewer than 8 years to complete.

But in one respect the complexity of most large systems today seems to be much greater, and that is in the complexity of the missions that the systems are asked to serve and in the number and diversity of users, supporters, and administrators of the systems. Indeed, it is often the increased complexity of external interfaces, more than internal system design complexity, that is the cause of extended development times and costs.

Software-intensive systems represent a special challenge because of the myriad of possible logic paths that can be woven through their codes. As Moore's law continues to drive down the size of computers and drive up their speed and capability, functionality that was once deeply embedded in the physical configuration of components has begun to emerge as software, enabling synergies among components that would have been unimaginable only a few years ago.

The successful design, manufacture, and operation of these complex systems demands an engineering discipline capable of comprehending and managing all of their components and their interactions, and that discipline is called systems engineering. Simply stated, SE is the translation of a user's needs into a definition of the system and its architecture through an iterative process that results in an effective system design.[5]

Systems engineering was born in the telecommunications industry of the 1940s and nurtured by the challenges of World War II, when project managers and chief engineers, with the assistance of key subsystem leads, oversaw the development of aircraft, ships, and other weapons systems. The post-World War II creation of more complex systems—for example, ballistic missiles and communication systems—led to the formalization of SE as an engineering discipline. The development teams, especially for large weapons systems, employed thousands of engineers and required the use of formal methods to integrate subsystems into useful and reliable systems.

Today the profession of SE is fairly well evolved. SE has experienced tremendous growth and recognition within the academic world and has a strong professional society advocate (the International Council on Systems Engineering, or INCOSE). Most importantly, SE has been recognized within industry as a profession that is critical to the development of complex systems.

As noted by Blanchard and Fabrycky,[6] systems engineering is good engineering with the following areas emphasized:

- A top-down approach is required, viewing the system as a whole. Although engineering activities in the past have very adequately covered the design of various system components, the necessary overview and an understand-

[5]A more rigorous and richer discussion of SE is found in Appendix C of this report.

[6]Benjamin Blanchard and Wolter Fabrycky, 2005, *Systems Engineering and Analysis* (4th Edition), Englewood Cliffs, N.J.: Prentice Hall.

ing of how these components effectively fit together has not always been present.

- A life cycle orientation is required, addressing all phases, including system design and development, production and/or construction, distribution operation, sustaining maintenance and support, and retirement and material phaseout. Emphasis in the past has been placed primarily on system design activities, with little consideration given to their impact on production, operations, support, and disposal.
- A better and more complete effort is required relative to the initial identification of system requirements, relating these requirements to specific design goals, the development of appropriate design criteria, and the follow-on analysis to ensure the effectiveness of early decision making in the design process.

A common illustration of the SE framework is the "Vee" model shown in Figure 1-1. The SE process begins at the upper left with the definition of user

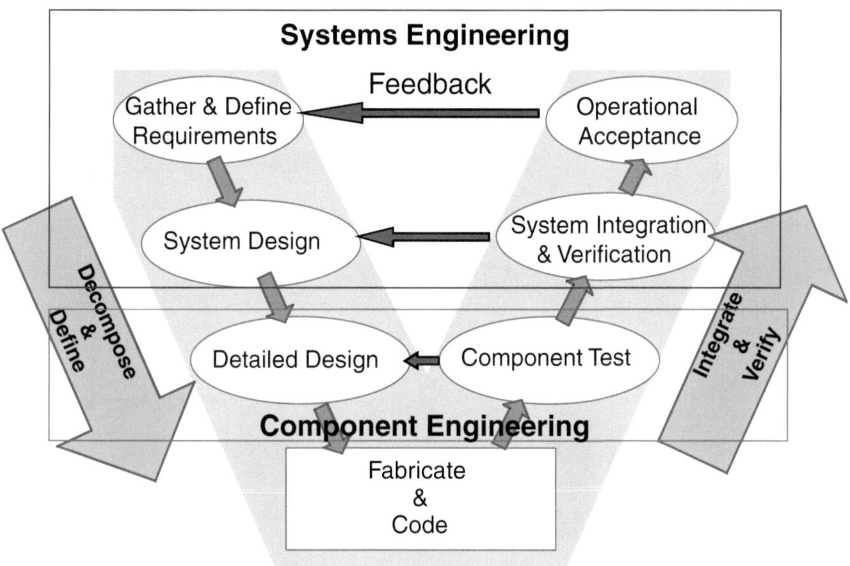

FIGURE 1-1 The "Vee" model of systems engineering. This model is generally attributed to the National Aeronautics and Space Administration, which in 1988 saw a benefit in bending the waterfall model into the "V" shape for software development. SOURCE: Modified from K. Forsberg and H. Mooz, 1992, "The Relationship of Systems Engineering to the Project Life Cycle," *Engineering Management Journal* 4(3):36-43. Copyright 1992 by IEEE. Reprinted by permission from IEEE.

requirements and of system concepts that meet those requirements. It continues down through system design and fabrication, then up through testing, integration, verification, and delivery of a product. Since SE encompasses the entire system life cycle, many SE diagrams continue to the right with segments representing system upgrades, maintenance, repair, and finally, disposal.

DEPARTMENT OF DEFENSE ACQUISITION PROCESS

Figure 1-2 provides an illustration of system development similar to that represented in Figure 1-1, but this time in the language of the DOD acquisition process.

Across the top of Figure 1-2 are the points at which important management decisions are made—Milestones A, B, and C. Concept development and refinement occur before Milestone A, and further technology development to flesh out the concept occurs before Milestone B. Only after Milestone B does a program become an enterprise with dedicated funding behind it. The nature of systems engineering changes significantly after Milestone B. Pre-Milestone A, systems

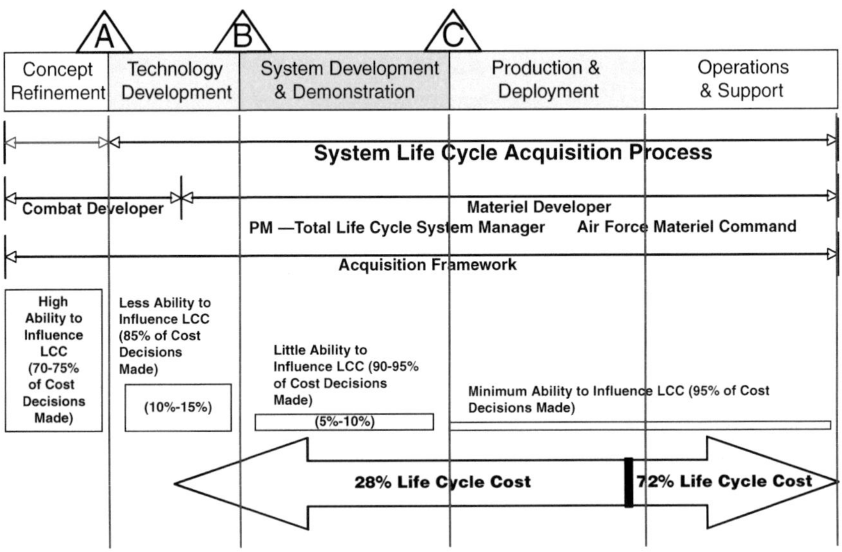

FIGURE 1-2 DOD life cycle acquisition process. Points A, B, and C at the top of the figure represent Milestones A, B, and C. LCC, life cycle cost. SOURCE: Richard Andrews, 2003, *An Overview of Acquisition Logistics,* Fort Belvoir, Va.: Defense Acquisition University. Available at http://www.afcea.org/events/pastevents/documents/Track4Session4AMCEmphasisonCustomerFocusedITInitiatives.ppt#364,12,Slide 12. Last accessed on November 20, 2007.

engineering involves translating the needs of the user into clearly stated key performance parameters (KPPs) and evolving the system concept and preliminary concept of operations (CONOPS) that satisfy these needs. After Milestone B, the emphasis shifts to the flowdown of requirements, interface management, performance prediction, system verification, and change control. The use of formal systems engineering practices throughout the life cycle of an acquisition program is critical to fielding the required system on time and within budget.

Importantly, Figure 1-2 shows that about three-quarters of the total system life cycle costs are influenced by decisions made before the end of the concept refinement phase at Milestone A, while about three-quarters of life cycle funds are not actually spent until after Milestone C. This means that while high-quality SE is necessary during the entire acquisition cycle, the application of SE to decisions made in the pre-Milestone A period is critical to avoiding (or at least minimizing) cost and schedule overruns later in the program.

HISTORY OF AIR FORCE DEVELOPMENT PLANNING

Prior to 1990, there were various organizations among the military service acquisition and development communities that focused on critical aspects of what is currently referred to as systems engineering. For example, the Air Force had within the Air Force Systems Command (AFSC) a structured organization whose mission was the front-end part of the total systems engineering process described above. "Strategy-to-task," a term invented by Lt. Gen. Glenn Kent[7] that describes the process encompassed by the left-hand portions of Figures 1-1 and 1-2, was addressed by such organizations.

The name given to these organizations was "Development Planning," or just "Planning." Their role was to employ various tools and techniques to define defense strategies, identify gaps in accomplishing those strategies, define concepts to address the gaps, use modeling and simulations or prototyping as ways to refine and test concepts, and provide early systems requirements to the systems developers for specific programs. Inherent in this role was the ability to understand the state of the art of the technical possibilities available from technology centers (laboratories, universities, industry, and so on), as well as to understand the needs of the user community (warfighters). These are all key attributes of a good pre-Milestone A systems engineering process. Successful programs discussed in Chapter 2 as "best practices" (e.g., C-5 and B-2) were originated during the "development planning" era.

Unfortunately, these planning organizations within the Air Force began to erode in the early 1990s, and at the same time the resources to support their

[7]Edward L. Warner III and Glenn A. Kent, 1984, *A Framework for Planning the Employment of Air Power in Theater War*, N-2038-AF, Santa Monica, Calif.: RAND; see also David E. Thaler, 1993, *Strategies to Tasks: A Framework for Linking Means and Ends*, MR-300-AF, Santa Monica, Calif.: RAND.

missions eroded. It is not clear whether the changes in the overall acquisition structure for the Air Force precipitated this erosion. However, the decreased role of the Air Force's acquisition command (originally AFSC, now the Air Force Materiel Command [AFMC]) in program execution contributed to this situation because there ceased to be a strong command-led focus on the function of development planning. Previously, the acquisition command ensured the availability of funding, manpower, and processes to support this front-end planning at every systems development center (aeronautics, electronics, weapons, and space). This role of the acquisition command headquarters also provided a sort of standards and evaluation of the processes and tools used by the various development planning organizations.

Today, there is renewed interest within the Air Force in strengthening the involvement of the acquisition command in the total acquisition process and program execution. As a part of this initiative, there is an opportunity to task the command to once again be the functional lead for development planning.

STATEMENT OF TASK AND COMMITTEE APPROACH

The Committee on Pre-Milestone A Systems Engineering was tasked to look at the role that SE can play during the defense acquisition life cycle in addressing the root causes of program failure. The original statement of task that had been developed with the sponsor[8] before the study began addressed the role of systems engineering in the full defense acquisition life cycle. During the committee's first meeting with the sponsor, it became apparent that, while the full acquisition life cycle was of interest, the sponsor was especially interested in the role that systems engineering could play during the pre-Milestone A and early phases of a program. The Air Force has concluded that many potential problems can be addressed by sound SE throughout the acquisition life cycle. However, currently there are few formal SE processes applied to Air Force development programs prior to the Milestone A review. The committee agreed to devote most of its attention during the study to the pre-Milestone A and early phases, and the sponsor and the committee agreed to revise the statement of task accordingly. See Box 1-1.

However, a limited broader discussion is necessary, because (1) systems engineering best practices and lessons learned apply through multiple program phases; (2) pre-Milestone A and early-phase systems engineering does not guarantee successful acquisition if poor decisions are made in subsequent pro-

[8]Terry Jaggers, the Deputy Assistant Secretary of the Air Force for Science, Technology, and Engineering, sponsored the study. In consultation with the National Research Council's Air Force Studies Board, he initiated the study and framed its terms, arranged for the Air Force to award and fund the study contract, and helped author the statement of task. He also collaborated with Mark Schaeffer, Director of Systems and Software Engineering in the Office of the Deputy Under Secretary of Defense for Acquisition and Technology. Mr. Schaeffer actively supported the study, essentially acting as an "unofficial" cosponsor.

BOX 1-1
Statement of Task

The National Research Council (NRC) will

A. Examine Air Force programs that were considered acceptable and unacceptable by DOD and assess the contribution of pre-Milestone A and early-phase systems engineering to the positive or negative development outcome. Milestone A is defined as the end of the Concept Refinement Phase and the beginning of the Technology Development Phase of the acquisition life cycle.

   1. From these examples, describe ways that pre-Milestone A and early-phase systems engineering were, or should have been, accomplished to produce successful results.

   2. Assess, describe, and when possible quantify the benefits of pre-Milestone A and early-phase systems engineering in successful programs, and the results of poorly executed systems engineering in terms of cost, schedule, and performance.

B. Determine the minimum level of pre-Milestone A and early-phase systems engineering required for program success, and current Air Force barriers to implementation, both on concepts leading to an Analysis of Alternatives (AoA) and for the post-AoA selected alternative.

C. Develop a framework/methodology for requirements and development organizations to use to ensure proper pre-Milestone A and early-phase system engineering is accomplished.

D. Discuss the positive effects expected to accrue across the remainder of the life cycle by properly accomplished systems engineering during the pre-Milestone A and early phases.

E. Discuss issues associated with adequacy and training of the entire workforce relevant to requirements, acquisition, and technology.

F. Recommend, in terms of law, policy, processes, and resources (tools, manpower, and funding) changes to enable and ensure the Air Force conducts adequate pre-Milestone A and early-phase systems engineering and the means for seamless transition from concept development through the genesis of a program office.

gram phases and, conversely, program failure post-Milestone A is not necessarily attributable to poor pre-Milestone A or early-phase systems engineering; and (3) the availability of data and proven models that could be used for analysis to causally link pre-Milestone A and early-phase systems engineering with program outcome is questionable.

To address the statement of task, the committee reviewed the literature and heard from many speakers involved in defense system acquisition programs.[9] The committee also drew on the expertise and extensive knowledge possessed by its members, who have had many years of personal experience both in defense acquisition programs and in the practice of SE.

The study sponsor did not provide the committee with a list of programs that are considered "successful" or "unsuccessful" by DOD. In addressing statement of task item A, the committee used its own judgment and examined a number of existing case studies of DOD programs in the literature and developed several new ones that it felt illustrate both successful and unsuccessful application of SE, focusing particularly on the pre-Milestone A and early phases. The committee notes, however, that many programs that are judged successful in retrospect were considered to be in trouble during their execution. Thus the perception of "successful" and "unsuccessful" programs can change with time and perspective. For purposes of this report, however, the committee does not believe that a program that requires far more time or money to develop than it would have in the 1960s and 1970s can be judged successful even if it ultimately meets its objectives. In a world in which the threats and technology are both evolving ever more rapidly, one cannot be satisfied with excessive deployment times and costs. The lessons learned from these case studies are summarized in Chapter 2.

The committee quickly determined the near impossibility of quantitatively isolating, testing, and proving direct causal links between pre-Milestone A and early-phase SE and later program cost, schedule, and performance outcomes. Many studies have searched for and proposed actions to address the root causes of the cost, schedule, and performance problems that seemingly have become the norm for current defense acquisition programs. Consistently, such studies have found that the causes and their effects are complex and interrelated.[10] The committee believes that high-quality pre-Milestone A and early-phase SE certainly contributes to later positive outcomes; however, available data did not allow the contribution of that SE to be reliably isolated from that of other factors such as requirements maturity and stability, funding stability, and domain knowledge of the development team. In that context, the committee addressed statement of task item A(2) qualitatively.

The committee addressed statement of task items B and C together by developing a checklist of items that constitute good SE practice during both the pre-Milestone A and pre-Milestone B periods. This checklist is presented in Chapter 4 (Box 4-1). As required in statement of task item B, the checklist items are divided into those that should be completed prior to the analysis of alterna-

---

[9]See Appendix B for a list of speakers and the presentations made to the committee.

[10]Ronald Kadish, Gerald Abbott, Frank Cappuccio, Richard Hawley, Paul Kern, and Donald Kozlowski, 2006, *Defense Acquisition Performance Assessment*. Available at http://www.acq.osd.mil/dapaproject/documents/DAPA-Report-web/DAPA-Report-web-feb21.pdf. Last accessed on April 2, 2007.

tives (AoA), and those that may be completed afterward. Associated with most checklist items are brief statements of the benefits that are expected to accrue throughout the program life cycle as a result of properly executing that item, as required in statement of task item D.

The discussion of issues associated with the systems engineering workforce and training, required under statement of task item E, is taken up in Chapter 3. That chapter provides a snapshot of the demographics of the current SE workforce as well as insights into current industry SE programs to help keep DOD programs on time and on budget.

The committee's policy recommendations on needed changes to law, processes, and resources, required under statement of task item F, are distributed throughout the report as they arise in the context of the discussion in various chapters. These recommendations are presented together in the Summary at the beginning of the report.

In preparing this report, the committee assumed that readers are generally familiar with SE and defense acquisition. The committee included brief definitions and descriptions where appropriate; however, it did not attempt to provide extensive tutorials on these subjects. Doing so would have been well beyond its charter and resources, and extensive literature and online resources already serve that purpose.[11]

[11]See, for example, the International Council on Systems Engineering at http://www.incose.org/practice/whatissystemseng.aspx, and the *Defense Acquisition Guide Book* at https://akss.dau.mil/dag/welcome.asp.

2

Relationship Between Systems Engineering and Program Outcome

INTRODUCTION

As discussed in Chapter 1, this committee places great importance on systems engineering (SE) processes and their proper application by domain experts throughout the entire acquisition cycle, but particularly in the earliest stages of programs, that is, pre-Milestone A. This chapter discusses the development history of a variety of past and ongoing programs, emphasizing the role of SE during pre-Milestone A and Milestone A-to-Milestone B time frames and deriving key lessons from the program outcomes.

The committee observed that programs which were successful in constructing a sound requirements baseline and financial/acquisition plan through the pre-Milestone A and Milestone A-to-Milestone B phase, using sound systems engineering processes, could succeed or fail. Those that failed were traced to poor post-Milestone B actions. However, those that had successful pre-Milestone A and Milestone A-to-Milestone B phases were the only programs that succeeded. The committee observed that there were no successful programs that entered into Milestone B without the sound fundamentals provided by the rigor and discipline of the systems engineering and financial/programmatic planning afforded by the pre-Milestone A and Milestone A-to-Milestone B processes. In this context, programs that succeeded were those that delivered their products within a reasonable margin of the original cost and schedule baseline. Programs that failed, in the committee's view, may have delivered successful products but were well outside the reasonable expectations of the original program and were only successful in delivering products after the addition of substantial unplanned funding and a substantial extension of the original schedule.

While the committee attempted to address the Milestone A and B issues in the case studies, it found very little information because of poor formal documentation regarding what happened in the Milestone A and B phases. Much of what is reported comes from the collective knowledge of committee members who were familiar with the programs. The point is that there is generally insufficient documented work done pre-Milestone A and B. The committee believed it important to include cases such as the C-5A to illustrate that good Milestone A and B work is not sufficient. It does not guarantee successful acquisition if poor decisions are made in source selection or during system design and development (SDD).

While it is not possible in most instances to draw from the cases a clear causal relationship to the individual findings and recommendations of this report, the committee found the case studies to be of value and expects that they will also provide useful information for the reader.

The important steps in the acquisition process are depicted as a continuous "thread" in Figure 2-1. Each segment of the thread is associated with a specific SE process. As Department of Defense (DOD) acquisition pulls the entire systems engineering thread, the thread can break at many different points for many different reasons. Thread breakage from perturbations in the SE processes tends to result in cost and schedule overruns and performance degradation.

The study schedule did not permit nor were the resources available to enable the committee to conduct in-depth, long-term studies of Air Force acquisition programs using a formal, rigorous, structured case study methodology. Instead, the committee had to use and to draw from published program data and formal

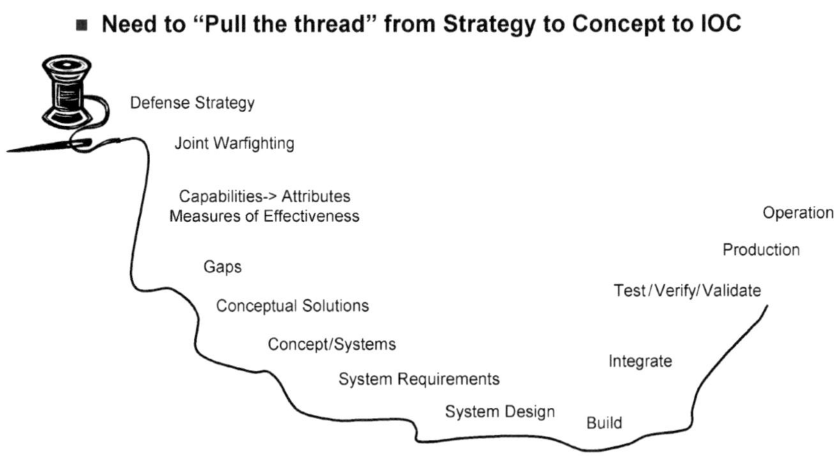

FIGURE 2-1 Systems engineering thread from defense strategy to system operation. IOC, initial operational capability. SOURCE: Contributed by committee member John Griffin.

work already done by others (such as the Air Force Institute of Technology's [AFIT's] published systems engineering case studies[1]), the presentations made to the committee on a sample of Air Force programs, and the knowledge and judgment of its members developed through long experience with such programs.

The committee examined the formal systems engineering case studies that had been completed by AFIT's Center for Systems Engineering. The AFIT case studies were based on the Friedman-Sage systems engineering case study methodology.[2] Each of the AFIT studies addressed the full program life cycle, including the early phases of a program (they were not specifically or strongly focused on pre-Milestone A, however).

For other programs the committee examined, the committee could not conduct analyses equivalent to the AFIT case studies. As the sponsor and committee anticipated, the committee did not find a wealth of data or proven models that could be used to strongly link a program's pre-Milestone A systems engineering effort with later program outcomes. Still, these programs yielded lessons learned that applied to the study charge.

The programs summarized in this chapter are these:

- Space Based Infrared Systems (SBIRS),
- Joint Direct Attack Munition (JDAM),
- Future Combat Systems (FCS),
- F-16 Fighting Falcon,
- Turbine engine development: fighter jet engine,
- Military Satellite Communications (MILSATCOM),
- C-5A, and
- B-2 Stealth Bomber.

These examples were chosen to illustrate the diversity of programs to which SE is applied: for example, from relatively simple systems such as JDAM to complex systems of systems such as FCS and MILSATCOM.

For brevity, program histories are summarized here; more detailed histories of the programs summarized by the committee can be found in the references provided in footnotes in this report. Other programs also considered in the committee's deliberations are mentioned in this chapter and, while not summarized herein, can be accessed through the published documents referenced in this report.

---

[1]Air Force Center for Systems Engineering Case Studies. Available at the website of the Air Force Institute of Technology at http://www.afit.edu/cse/cases.cfm. Last accessed on November 20, 2007.

[2]See G. Friedman and A.P. Sage, 2005, "Case Studies of Systems Engineering and Management in Systems Acquisition," *Systems Engineering* 7(1):84-97. The "Friedman-Sage" (F-S) SE case study methodology was used in the generation of the AFIT case studies; however, the AFIT case studies went well beyond the F-S methodology in the interview process and so on.

PROGRAM SYNOPSES

This section highlights the application (or lack thereof) of SE to past and ongoing acquisition programs and their performance during various phases of their life cycles.

Space Based Infrared Systems (SBIRS) Program

SBIRS is designed to provide tracking and targeting capabilities for missile warning, missile defense, and technical intelligence. SBIRS is a successor to the highly successful Defense Support Program (DSP) satellite system, which performs the missions of early warning of strategic ballistic-missile launches, detection and reporting of tactical missile launches in theaters of interest, and, as secondary missions, detection of space launches and nuclear detonations.

The histories of the SBIRS Program and DSP present a stark picture of how procurement time lines and costs have ballooned over the past 40 years. The contractor for the initial DSP development was selected in 1965, and a successful initial operational capability (IOC) was achieved just 5 years later. Since that time, DSP has been improved many times and remains a valuable part of U.S. space assets to this day. The successor to this program, now called SBIRS, was awarded for full-scale development in 1996. Since that time, numerous schedule slips and cost overruns have been announced, and the first launch is now scheduled for 2009. During the SBIRS briefing to the committee, the presenter said that the program's minimum goal is to produce a first flight article that is "at least somewhat better" than DSP.[3]

The SBIRS program has received much attention due to program execution issues. The program experienced Nunn-McCurdy breaches[4] in 2002 and 2005, at which times it had to be recertified. The program had two additional Nunn-McCurdy breaches of 15 percent cost increase, but did not have to be recertified. The SBIRS program has completed Increment 1 and the delivery of its first two hosted sensors and is currently working to deliver its first geosynchronous Earth orbit (GEO) satellite; however, this milestone is projected to be reached extremely late relative to original program plans and will have considerably exceeded original cost projections.

Systems Engineering Lessons

The committee believes that the primary factors in this situation are as follows:

---

[3]Col. Randall Weidenheimer, 2007, "Space Based Infrared Systems Wing," Space and Missile Systems Center, USAF, presentation to the committee, January 31, 2007.

[4]A Nunn-McCurdy breach occurs when the program acquisition unit cost (PAUC) or average procurement unit cost (APUC) exceeds the baseline value by 25 percent or more.

- The internal complexity of the SBIRS spacecraft, including complex focal plane arrays and the need for concomitant signal processing algorithms, is much more complex than that of the DSP spacecraft.
- The external user interfaces for SBIRS are far more complex than those for the earlier DSP owing to requirements of the missile defense community, tactical users, and the legacy early-warning role.
- There was a high degree of requirements instability after Milestone B (or its equivalent), driven by shifting missile defense strategies and tactical requirements changes.
- Full-scale development was undertaken while the sensor technology initially planned was not mature.
- There was a high reliance on new flight software.
- There was inadequate oversight and accountability on the part of both the Air Force and the contractor.

Each of these systems engineering issues was a significant contributor to the SBIRS Program's failure to execute. However, since the issues of technology readiness, acquisition process, and requirements stability have been dealt with thoroughly in prior reviews of the SBIRS Program, the committee focuses here on the software issues, which have been overlooked elsewhere.

The SBIRS Program represented a several-fold increase in the scope and scale of flight software relative to earlier Air Force programs and a manyfold increase over legacy systems in the SBIRS mission area. However, software systems engineering received insufficient emphasis and was not effective in early program formulation. Software and mission domain experience, in both the government and contractor teams, was inadequate to prevent or mitigate the impacts of inappropriate flowdown of requirements to the software, overly optimistic projections of software production rates, and inappropriate phasing of software development and testing with the system-level integration and test program. In fact, flight software still poses a risk to the execution of the remaining work in SBIRS. This risk is less related to how many lines of code can be written and tested in a given period of time than it is about whether or not the software system engineering was sufficient to produce the robust and resilient flight software system that will be needed over the current and succeeding phases of this complex program.

Several other programs have recently experienced program execution issues late in the development phase as a result of failures to employ robust software systems engineering principles in the pre-Milestone A phase. These program execution issues are frequently not attributable to pure software causes; instead, they tend to occur (perhaps more insidiously) in programs with complex hardware/software interfaces or dependencies (such as with SBIRS). Another way to characterize the problem is that there was insufficient coupling between the disciplines of systems engineering and software engineering early in the program.[5] As a result of these problems, lessons are being learned, progress is being made, and improvements in practices and methodologies are being employed in some current pre-Milestone A programs (discussed below).

Traditionally, there has been fairly loose contact between the disciplines of systems engineering and software engineering, both in academia and in practice. Typically, software engineering has contributed more to the formation of the discipline of systems engineering than the reverse, primarily because systems engineering as an academic discipline is a younger field. However, there is a notable trend lately of these two fields coming together, at least in academia. For example, the University of Southern California's Systems Engineering Center recently merged into its Center for Software Engineering to form the Center for Systems and Software Engineering.

These trends in academia are beginning to carry over into practice, leading to better up-front treatment of software systems engineering in the pre-Milestone A phase. A decade ago, when programs such as SBIRS were in their formative phases, program managers and chief engineers might have readily ignored the question of the feasibility of accomplishing stressing system requirements using software. Indeed, it had been common for software to be the fix-all for any design requirement that was deemed "too hard" for the hardware design—"We'll just do it in software." Today, the hardware/software trade space is more typically dealt with using the same level of rigor as is used in other system design trades. This increased contact between the disciplines of systems engineering and software engineering will allow early input into the requirements allocation process and trade space analysis by informed software specialists, thereby avoiding the later program execution pitfalls that result from having expected all of the hard problems to be handled by the software.

Specifically, the lessons learned from the SBIRS Program are being taken to heart in several more recent programs. For example, in the earliest phases of the Space Radar Program, the system program director chose to create a software division within the government program office, along with divisions focusing on the space vehicle, ground system, and systems engineering. With this organization, the program director sought to ensure that all pertinent systems engineering trade analyses and programmatic decisions would be vetted through all competing needs of the program, and that the final decision options would be presented with a full characterization of the benefits and impacts associated with each discipline area, including software. Surely, the Space Radar Program will face its share of technical and programmatic challenges, but it will be informative to see to what degree potential software systems issues have been avoided or their impacts mitigated by this organizational approach to software systems engineering in the formative phase of the program.

[5]Robert N. Charette, 2005, "Why Software Fails," *IEEE Spectrum* 42(9):42-49.

The committee notes here that one of the items on its systems engineering checklist (see Chapter 4) is to assess the methodology that the program has chosen to integrate systems engineering and software systems engineering.

SBIRS and DOD Acquisition Reform

The SBIRS acquisition was undertaken by the Air Force during a period in which "acquisition reform" was being strongly pursued. One of the premises of acquisition reform was to reduce life cycle cost through reduced government program office size and responsibility. The DOD began a program to reduce the size of the acquisition workforce over a period of several years. The Congress participated with direction in the Fiscal Year (FY) 1996, FY 1997, FY 1998, and FY 1999 Defense Authorization Acts to reduce the acquisition workforce.[6]

A consequence of reducing the number of government acquisition personnel was the outsourcing of more acquisition work to industry. The SBIRS Program elected to delegate total system performance responsibility (TSPR) to the prime contractor. Assignment of TSPR to the prime contractor for systems of this scope and complexity is inappropriate if it leaves the government in a position of seeking insight into program execution and decisions, rather than assuring accountability through active oversight. While there was nothing in the definition of TSPR that required no government involvement or that specified no government accountability, the implementation of TSPR on SBIRS was taken to an extreme.[7] A symptom of the flawed acquisition approach was the abrogation of government accountability for many of the pre-Milestone A systems engineering processes and products.

Some acquisition reform initiatives (many in the Air Force directed under the heading of "Lightning Bolts"[8]) were sent to the field for acquisition programs without a complete understanding of their consequences. While some seemed sensible in theory, many were applied inconsistently, resulting in unintended consequences. The committee notes that it would be a repeat of past failures to invent a series of SE reforms and mandate them in a "one size fits all" fashion without assessment and tailoring to the situation. Many of these same approaches worked fine in other programs (see the following case of JDAM). These points are discussed further in the systems engineering checklist presented in Chapter 4.

---

[6]See the Chapter 3 section entitled "Congressional Actions to Cut DOD Acquisition Workforce" for additional discussion of these acts.

[7]The issue here is the word "total." The government program office cannot give up total responsibility for cost and performance.

[8]"Lightning Bolts" is the title given to a series of acquisition reform initiatives launched in 1994 by then Principal Deputy Assistant Secretary for Acquisition and Management for the Air Force Darleen A. Druyun. The initiatives, which included establishing a centralized acquisition support team to scrub all proposals over $10 million, developing a new System Program Office model, and reducing the number of military specifications and standards, were intended to make the acquisition and sustainment processes for the Air Force better, faster, and cheaper.

It would be a gross oversimplification to attribute all SBIRS program execution issues solely to acquisition reform. Likewise, the program facts do not support faulting systems engineering for all SBIRS execution issues. Nevertheless, poor SE did contribute to program troubles encountered thus far and might yet be a major contributor to execution issues with the remaining effort, particularly in the area of software systems engineering. The SBIRS case provides impetus to assess the level and quality of the integration of systems engineering and software systems engineering in ongoing programs.

Joint Direct Attack Munition (JDAM) Program

Description

JDAM was initiated in late FY 1991 and had its roots in operation Desert Storm. It was during that conflict that military leaders realized the need for all-weather, extremely accurate bombs capable of being dropped from a number of aircraft platforms. The military arsenals were filled with hundreds of thousands of "dumb" gravity bombs. The military wanted to turn these unaided bombs into "smart" bombs using a strap-on kit. The kit would use Global Positioning System (GPS) satellite-guided signals and computer technology to deliver the bomb within 13 meters of its target, regardless of environmental conditions such as storms, darkness, and high winds.

The JDAM Program operated in the same acquisition reform environment that SBIRS operated in, but the team did not select a TSPR approach, and the team's perseverance paid off. The JDAM Program is a success by every measure: the JDAM team's final proposal included an average unit production price (AUPP) between \$14,000 and \$15,000—down from an original cost target of \$40,000 and an original cost estimate of \$68,000. The JDAM team reduced its research and development costs from \$380 million to \$310 million and shortened the development program length from 46 months to 30 months. The total procurement cycle length was reduced from 15 years to 10 years. Military performance has continued to improve (e.g., product improvements have allowed the JDAM Program to contractually tighten performance accuracy from the original 13 meters to 5 meters). By February 2007, more than 150,000 JDAMs had been delivered to U.S. and international customers and had been integrated onto 10 U.S. and 5 international aircraft platforms. JDAM is a truly operationally effective system.

There were many other innovative acquisition and systems engineering initiatives employed by both the government (system program office [SPO] and user) and the industry teams.[9]

---

[9]For more details, see C. Ingols and L. Brem, 1998, *Implementing Acquisition Reform: A Case Study on Joint Direct Attack Munitions (JDAM)*, Washington, D.C.: Defense Systems Management College. Available at http://www.acqnet.gov/comp/seven_steps/library/JDAMsuccess.pdf. Last accessed on September 18, 2007.

Systems Engineering Lessons

The success of the JDAM Program was the result of many individual applications of effective systems engineering and program management principles. Primary among these were that (1) the requirements were clearly defined prior to development and remained stable throughout the development phase, (2) the technology used was mature, and (3) the external interface complexity was well managed.

Also of importance was the effective teaming between the government and contractors to form accountable Integrated Product Teams (IPTs). The IPT structure within the JDAM Program instilled a motivation to all team members to engage in and contribute to the success of the program. Another key ingredient was the up-front focus on affordability. This was facilitated by contract provisions allowing the contractor's profitability to increase as a result of cost reductions accrued through innovative technical and management approaches (including the selective use of commercial practices and specifications).

Future Combat Systems (FCS)

Description

The Army's Future Combat Systems (FCS) is, without question, the largest, most complex program in Army history. It is literally a "system of systems of systems." Figure 2-2, from an Army briefing on FCS to the National Defense Industries Association, illustrates this complexity. The overarching objective of the program is to develop a lighter, more lethal force that could be deployed far more rapidly than the heavy forces that are a major proportion of the Army force structure today. To achieve the overwhelming lethality that is envisioned, the concept of network-centric warfare is the underlying theme. It is expected that, via an integrated, robust, self-adapting, self-healing network, pertinent situational data will be available to all leaders at all echelons of the Army ground force (i.e., they will have accurate situational awareness). This will support operations well within the enemy's decision/awareness cycle and achieve increased lethality.

These concepts were evolved by the Defense Advanced Research Projects Agency (DARPA) in cooperation with the Army and taken through early concept development. At that stage, the Army "constructed" the FCS Program. The decision was made to take the concepts as defined at that time and, via a competitive bidding process, select a lead systems integrator (LSI) contractor to pull together the doctrinal concepts and codify them into a system of systems architecture. It was further decided to use Other Transactions Authority (OTA) as the program management vehicle, rather than the traditional milestone process. The idea was that early capabilities/requirements analysis leading to the systems of systems architecture and eventually to systems specifications could progress much faster without the layered milestone decision authority process. It was not envisioned

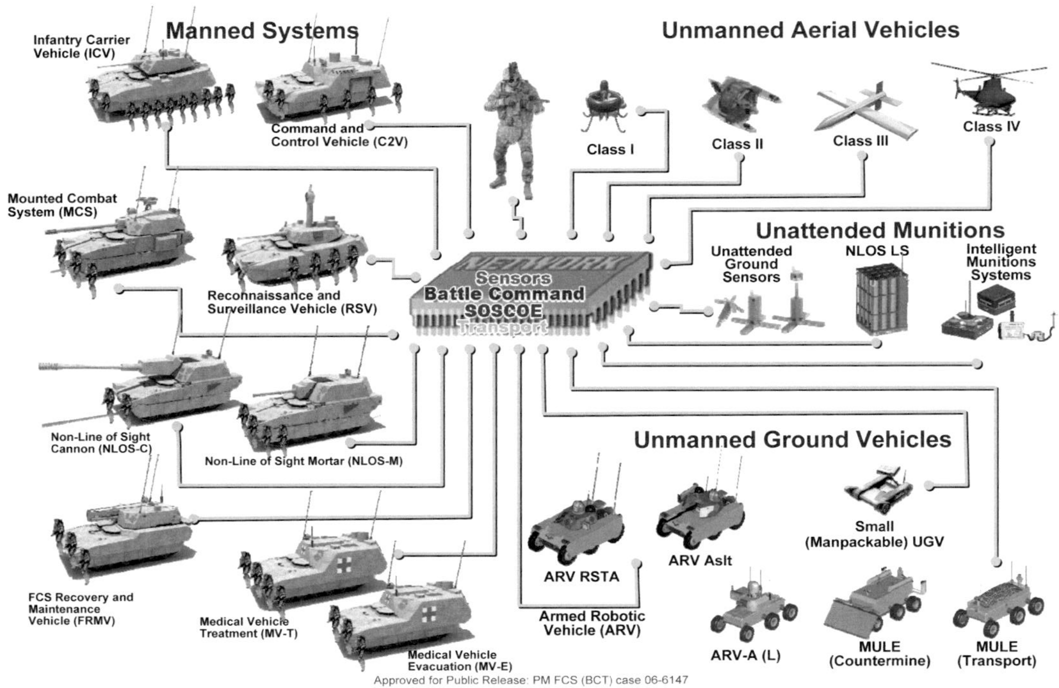

FIGURE 2-2 Conceptual illustration of the Future Combat Systems Program and its complexity as a system of systems. NOTE: SOSCOE, System of Systems Common Operating Environment. SOURCE: D. Emery, D. Bassett, and M. Schnaidt, "The Net-Centric Foxhole: Perspectives from Army Future Combat Systems," presentation, October 25, 2006. San Diego, Calif.: National Defense Industrial Association 9th Annual Systems Engineering Conference. Available at http://www.dtic.mil/ndia/2006systems/Wednesday/emery.pdf. Last accessed on November 20, 2007.

that less rigor would or should be employed in the early capabilities/requirements determinations nor in the translation of capabilities/requirements to systems architecture and thence to systems specifications.

Significant amounts of time, effort, and dollars were spent in platform and sensor technologies, in parallel with trying to define the integrating network. The LSI seemed to recognize this difficulty early on and set about developing an integrating computer hardware/software system entitled System of Systems Common Operating Environment (SOSCOE). Without some capability such as SOSCOE, there would have been little chance that the various platforms could ever be integrated. In parallel with SOSCOE, the network architecture began to be addressed. The network architecture early concept and definition work was constrained, however, by a mandate from the government that its central core would be the Joint Tactical Radio Systems (JTRS) waveform architecture. While on the surface this did not appear to be a major obstacle, it soon became a problem in that the JTRS program itself was having technical and schedule problems and became a gating item in defining the network operating concepts, architecture, and topology.

In summary, the activities that might be associated with normal pre-Milestone A work just did not coalesce. The program, under OTA, entered into full-scale development with many unresolved issues on systems capabilities/requirements and architecture. Costs were rising well above the original estimates, systems integration problems were growing, and the schedule was slipping well beyond the original estimates. Meanwhile, Congress was expressing real skepticism. As a result, the Secretary of the Army ordered that the program be restructured into the DOD-approved milestone process.

Systems Engineering Lessons

The FCS Program violates many of the precepts that the committee believes are important to success in the conventional system-development process. For example, the prime contract for FCS was let while requirements were very much still being traded off against desired capabilities. The Army has had difficulty stabilizing requirements changes between Milestone B and IOC, perhaps due to insufficient systems engineering in the early phases of program planning.

A preliminary observation based on this case concerns the need for rigor in those activities associated with defining capabilities and requirements and overall systems architecture, independent of the formality of DOD milestone definitions, before entry to the SDD phase is authorized. It should be recognized that, as a potential program progresses through its early stages, better insight into technical feasibility and evolving needs will sometimes necessitate changing the requirements. If the changes become substantial, the program should be recycled through concept development before going ahead with full-scale SDD. A firm conclusion requires more analysis. The full suite of essential requirements should be developed at Milestone A and be part of the reporting requirements at each of the program's major reviews. Once SDD is authorized, a program should be under fairly rigorous configuration management, not only for the hardware and software but for the requirements, to assess both creep and stability. Tracking progress against original requirements should be a checklist item for reviews. This may be integrated into the key performance parameter (KPP) process.

While the issues identified above would raise serious concerns in a conventional acquisition program, the committee notes that FCS is not a conventional development process. It is an LSI-managed procurement in which an LSI contractor (distinct from the development contractors) oversees the entire system engineering process. In fact, this process is set up deliberately to disregard some of the elements of good system development that the committee delineates in the SE checklist in Chapter 4. However, the ability to deal with such complexity and shifting design requirements is considered a strength of the LSI process. The committee does not know how this approach will turn out, as the program is still very much in the middle of the development process. If it is successful in delivering the desired transformational combat capabilities in a reasonable time and for a reasonable cost, it will serve as a model for future systems of systems procurements and will add new dimensions to the systems engineering art.

F-16 Fighting Falcon

Description

The F-16 has a reputation for success spanning more than three decades. There are many accounts of how this legendary fighter was developed and of the issues that the program faced as it evolved over the years. One of these, an article by Richard P. Hallion entitled "A Troubling Past: Air Force Fighter Acquisition Since 1945,"[10] is quoted in part below.

> Given how suitable the F-15 would ultimately prove to be for both the air superiority and air-to-ground roles, it is somewhat ironic that in 1968 (fearful that the Mach 2+ F-15 would turn out to be just another big, fast sled) Boyd, Spray, and the others began arguing for a highly agile, single-engine, and less-than-Mach 2 "austere" fighter, the so-called F-XX. They were unsuccessful in getting the Air Staff to redirect the F-15 program again—a wise decision on the part of the Air Force. Instead, the climate of thought that they proposed with the F-XX germinated at the end of the summer of 1971 in the so-called lightweight fighter program. The LWF program received a significant boost by a dramatic redirection of defense acquisition in June 1970, when then-president Richard M. Nixon's "Blue Ribbon Defense Panel" recommended ending so-called total

[10]Richard P. Hallion, 1990, "A Troubling Past: Air Force Fighter Acquisition Since 1945," *Airpower Journal* 9(4):4-23.

package procurement and returning to competitive prototyping, something that had been abandoned since the late 1950s.*

Ultimately this interest spawned a competitive fly-off between the General Dynamics YF-16 and the Northrop YF-17, and out of this fly-off came both the F-16 and F-18 airplanes. Although ostensibly intended for technology demonstration, there was little doubt that the "winning" aircraft would have an excellent chance for full-scale production. In mid-January 1975, the Air Force declared the YF-16 the winner, awarding a contract for full-scale development. The first F-16A, which was a slightly larger and more refined aircraft than the YF-16 demonstrator, flew in December 1976. The Air Force activated its first F-16 squadron in January 1979, roughly a decade from the time the fighter mafia initially called for its development. Widespread foreign sales followed. (The YF-16/YF-17 competition was a win-win situation for both contestants, for the losing YF-17 was subsequently adopted, in greatly modified form, as the basis for the McDonnell Douglas F/A-18. Mirroring pilot opinion of the F-15 and F-16, naval aviators generally were enthusiastic over its performance.)**

Unlike the F-15, the F-16 was a true fly-by-wire aircraft, using three computers constantly "voting" on each other's performance to maintain control of what was basically an unstable airplane. The F-16 thus possessed superlative maneuverability, really making it a six-and-one-half-generation airplane, demonstrating performance only now being approached by foreign designs such as the Soviet MiG-29, the European fighter aircraft (EFA), Israeli Lavi, French Rafale, and Swedish Gripen. It is worth noting that going beyond the original air superiority intentions of its parents, the Air Force acquired the F-16 as a dual-role air-to-air and air-to-ground fighter-bomber. By acquiring it, the Air Force intended to

*Neufeld, Jacob. "The F-15 Eagle: Origins and Development, 1964–1972." Air Power History 48, no. 1 (Spring 2001):4–21; M.B. Rothman. "Aerospace Weapon System Acquisition Milestones: A Data Base." Rand Corp.: N-2599-ACQ. October 1987.

**Deborah L. Gable, "Acquisition of the F-16 Fighting Falcon (1972-1980)," Report 87-0900 (Maxwell AFB, Ala.: Air Command and Staff College, 1987), is a useful survey, copy in the history files at Headquarters Air Force Systems Command; Rothman, 62-65. Like the F-15, the F-16 proved infinitely more tractable than its century-series forebears. Further, I have benefited from conversations regarding the F-16's flight control system with three noted test pilots who shepherded the plane from its YF-16 phase into full-scale development and on into operational service: Phillip Oestricher; Col Robert C. Ettinger, USAF, Retired; and Col David Milam, USAF. See also Spangenberg's VFAX/ACF program memorandum. All aircraft tend to have quirks and faults of one sort or another. Unfortunately, the very earliest production models of the F/A-18 have experienced significant problems with structural cracks, and this condition may limit the service lives of some 61 airplanes. See Barbara Amouyal and Robert Holzer, "Structural Flaws Halve Life of Early F/A-18 Hornets," Defense News, 20 November 1989, 3. Somewhat balancing this is that as a combat airplane, the Hornet has been very impressive to its flight and ground crews from performance, reliability, and maintainability standpoints. The Navy is planning to procure a total of 1,157 Hornets.

complement the more expensive and capable F-15 carrying a mix of medium- and short-range air-to-air missiles with a cheaper swing-fighter carrying Sidewinders that could assist in winning the air battle, and then fight air-land war. It is the F-16's multimission capabilities that subsequently resulted in orders for 3,000 of this type aircraft, placing it among the most successful of postwar jet fighters.

Systems Engineering Lessons

The F-16 featured many innovations in the application of engineering and management concepts, but fundamentally the advantages that this aircraft possesses reflect the shrewd application of available technology. Planners with extensive domain expertise were able to anticipate future warfighting environments, understand the systems acquisition process, and comprehend the state of technology to meet the needs.

A major lesson learned is that the F-16 Program applied sound systems engineering continuously through an evolutionary block upgrade of the weapon system. This has resulted in the evolution of the weapon system from a basic air-to-air fighter to one of the most sophisticated air-to-air and air-to-ground weapon systems in the world today. This example shows that SE is a life cycle effort and should be planned, programmed, and executed as such.

The implications of such a life cycle approach are significant and hold key lessons and introduce difficult questions for application for future weapons systems. The F-16s in the first block produced were very basic in their functional capability, providing for the integration of key technologies in the basic platform. Follow-on blocks would improve the combat capability in a predictable and stable engineering and management environment. The development time line, therefore, was shorter for the individual blocks and within a management and oversight time frame. This approach reduced the threat of instability in requirements and cost because the expectations were not only measurable but near term. The key enabler here was the ability and discipline to produce the first iteration of the platform with basic functionality—something not normally accepted in today's process. SE processes must address a key question pre- and post-Milestone A: Can an initial capability be achieved within the time that the key program leaders are expected to remain engaged in their current jobs (normally less than 5 years or so after Milestone B)? If this is not possible for a complex major development program, can critical subsystems, or at least a key subset, be demonstrated within that time frame? The SE process should provide this alternative for decision makers. The F-16 experience suggests that it can lead to very successful outcomes.

Fighter Jet Engine Program

Description

Aircraft engine systems offer an illuminating case in point to demonstrate the value of systems engineering processes to incorporate sustainability from the earliest phases of a program.[11] The engine systems referred to here are designated EG10-1, EG10-2, EG10-5, EG10-9. Collectively, these four engine systems are called the EG10 engine family. The original EG10-1 engine first qualified for use by the USAF in October 1973. Since that time, the contractor has improved on the design of the engine; its latest variant is the EG10-9 engine.

In 1977, the Air Force organized the Propulsion System Program Office (SPO) in order to oversee the development of the EG10 fighter engine and its rival engine system that was being developed in parallel by another manufacturer. The Propulsion SPO was responsible for ensuring that the engine systems fulfilled the Air Force's performance and sustainability needs. This organization of engineers and managers was also responsible for negotiating contracts between the Air Force and the engine manufacturers while assisting the manufacturers in solving problems that occurred during development, such as system integration issues and cost overruns. In many ways, the purpose of the SPO was to ensure that the engine's development problems did not recur with the new engine system.

The majority of SPO personnel were already veterans of previous engine system development projects and were already well acquainted with the many problems that an engine development program faces. This previous experience among team members was one of the most valuable resources in the success of the engine development program.

The SPO not only interacted with the original equipment manufacturers (OEMs) but also with the warfighters who would eventually make use of the new engine's capabilities. The SPO was responsible for surveying the warfighters in order to determine what they needed from the new engine in terms of maintainability and performance. Some SPO personnel had previously served with operational fighter squadrons as technicians and were able to bring firsthand knowledge of difficulties faced on the flight line. The SPO maintained a close relationship with the fighter squadrons to ensure that the warfighters' interests were communicated to the OEMs.

---

[11]This section is based on work done by committee member Wesley L. Harris, Massachusetts Institute of Technology (MIT), on case studies conducted on a series of Air Force fighter jet engine systems. Consistent with the bilateral agreement between the major American aerospace manufacturer and MIT, the manufacturer is anonymous and the engine systems studied are given the aliases "EG10-1," "EG10-2," and so on, collectively called the EG10 engine family. See "Sustainment Measures for Fighter Jet Engines" by Spencer L. Lewis and Wesley Harris, Society of Automotive Engineers, Inc., available at http://dspace.mit.edu/bitstream/1721.1/723/1/07_12_2001_Sustainment.pdf. Last accessed on June 29, 2007.

In the 1970s, Air Force Propulsion SPO managers instituted ground tests that more accurately simulated the operation of an engine throughout its lifetime. Designers used information on what throttle settings would be utilized throughout the life of the engine, given various expected mission types (e.g., air-to-ground missions, air-to-air missions, and escort missions). They then transformed a composite of these throttle settings into a simulation of the throttle settings at which the engine would be expected to perform.

The EG10-5 eliminated many of the problems suffered by the EG10-1 and EG10-2 engines, such as in-flight stall stagnations and problems with the fuel control system. Additionally, because efforts were made to understand better how the engine would be employed in the field, the squadron owners of the engines were more satisfied with how they were able to maintain the system.

The Air Force's pro-sustainment policies did not cease upon completion of the EG10-5 project. The OEMs wished to continue producing engines for the F-15 and F-16 aircraft, and in order to do that, they had to continually improve the sustainability of their engines. Owing to the Air Force's strong emphasis on sustainability, the two corporations installed a number of new features (such as electronic monitoring systems) making their systems more sustainable. It was in the midst of these innovations that the EG10-9 engine was designed.

Systems Engineering Lessons

Systems engineering contributed significantly to the improvements in reliability and maintainability of the evolving engines in the EG10 family of engines. These contributions may be classified as policy, technology, and process and tool development. Systems engineering thinking was used to develop an effective policy transition, managed by the Propulsion SPO, from a "nonsustainment" ideology to a "performance-with-sustainment" requirement. Technology contributions that were based on systems engineering included computer-aided design, modularity, electronic engine controls, and computer-aided logistics. In the area of processes and tools, systems engineering thinking enabled the development and use of IPTs and accelerated mission testing. These tools were captured in the transition from sequential engineering to concurrent engineering in the design and manufacture of jet engines.

The history of the fighter engine development programs shows that designing early for sustainment and performance results in more affordable and agile life cycle options, and a greater flexibility for technical upgrades throughout the operational lifetime of a system. Perhaps most importantly, however, it illustrates the value of having an experienced, domain-knowledgeable organization within the government (in this case, the Propulsion SPO) to manage acquisition programs.

Military Satellite Communications (MILSATCOM)

Description

One of DOD's most crucial goals is to create an interoperable system of systems aimed toward enhanced command, control, communications, computers, and intelligence. The mission of the MILSATCOM Program is to provide global, space-based communications capabilities supporting DOD and other government agency missions.

MILSATCOM is itself a system of systems, providing narrowband, protected, wideband, and network communications capabilities to a wide range of military users. MILSATCOM must interface with a multitude of external stakeholders, some of which control key architectural specifications, or even performance requirements for the system (e.g., information assurance).

The first mission area is the narrowband satellite communications (SATCOM) area that includes the UHF Follow-On (UFO) system and Mobile User Objective System (MUOS). Narrowband provides reliable service to mobile users.

The wideband mission area includes the Defense Satellite Communications System (DSCS), its follow-on Wideband Gapfiller System (WGS), and Global Broadcast Service (GBS). Wideband provides broadcast service, similar to DirecTV, and high-capacity data pipes.

The protected mission area is supported by Milstar, advanced extremely high frequency (AEHF) system, and Polar MILSATCOM. Protected SATCOM provides highly secure and survivable communications. The Transformational Satellite Communications System (TSAT) is a next-generation MILSATCOM system, following WGS and AEHF. It will support both the protected and wideband mission areas.

In addition to the satellite systems, MILSATCOM includes a terminals segment. The different SATCOM terminals communicate with one or more of the satellite systems. Examples include the Milstar Command Post Terminal, Secure Mobile Anti-Jam Reliable Tactical Terminal (SMART-T), Family of Advanced Beyond-Line-of-Sight Terminals (FAB-T), DSCS and GBS Terminals, Ground Multi-band Terminal (GMT), Airborne Integrated Terminal (AIT), Navy Multiband Terminal (NMT), Warfighter Information Network-Tactical (WIN-T), High Capacity Communications Capability (HC3), and Multi-Band Multi-Mode Radio (MBMMR).

Historically, the individual MILSATCOM systems were handled in a piecemeal fashion—each system focused only on what it needed to be individually successful, not considering how it might contribute to the broader MILSATCOM program. Each of the satellite systems, while being headquartered at the Air Force's Space and Missile Systems Center, had its own user community, its own program director, program office, budget, schedule, and so on.

In recent years, a combination of increasing integration of military operations and a dramatic Air Force drawdown of technical and program management talent have forced organizational change in the management of MILSATCOM programs. Today, the MILSATCOM Program is being handled as an enterprise under the MILSATCOM Joint Program Office (MJPO). Within the MJPO, each system is assessed not only on its own merits, but also on how it contributes to the various MILSATCOM mission areas (protected, wideband, narrowband SATCOM) and how it can contribute to meeting some or all of the needs of a particular user community (for example, supporting intelligence, surveillance, and reconnaissance data relay).

The MJPO has taken several measures to reduce program risks and increase the probability of success in delivering MILSATCOM capabilities. These efforts have focused primarily on the integration impacts in four areas: across the breadth of individual programs, in the transition to operations, in the interactions among programs, and external influences on programs. As a result of interconnections between various MILSATCOM systems, the MJPO has also recognized the interdependencies among programs. The MJPO has started several initiatives to reduce the risks of any program's adversely impacting others. These initiatives include the stand-up of the MILSATCOM chief engineer and the System of Systems Engineering, Architecture and Integration (SSEA&I) group; the MJPO Configuration Board (MCB); weekly senior-level functional reviews and discussions; and the MILSATCOM Integration Master Schedule (IMS). The MJPO created the MCB to track and manage the integration of interfaces, specifications and standards, and program efforts. This executive-level board, composed of MJPO program managers, is responsible for advising the MCB chair (MJPO director or deputy) on proposed contract change actions, and manages all MJPO configuration baselines throughout sustainment and any changes that have a cost impact.

Despite the move toward unified program management represented by MJPO, in recent years the development of the MILSATCOM systems has been plagued with delays and cost growth that are a legacy of the earlier, fragmented management structure. Examples are the AEHF and WGS systems.

Following the failure of Milstar Flight 3, AEHF was accelerated to complete worldwide protected SATCOM coverage. This decision forced significant adjustments to AEHF Flight 1, including a transition to an operations plan that is less than optimal. These changes resulted in the addition of a requirement that AEHF be backward-compatible with Milstar. Extra attention is being given to the AEHF system development effort to address issues on Milstar backward compatibility, operations transition, cost overrun projections, and sustainment. The AEHF system procurement effort has had to provide a Nunn-McCurdy notification to Congress owing to late delivery of government-furnished products (especially for performing cryptographic functions) and replacement of critical electronic parts, causing a cost breach greater than 15 percent. AEHF had to replan the procurement effort, delaying delivery of the first satellite, and is now subject to increased scrutiny within the DOD and in Congress.

The WGS system was contracted during the so-called Acquisition Reform era, in which the government expected "commercial best practices" to be applied.[12] The program viewed the WGS satellites as a mature product requiring little oversight. As it turned out, the large number of program interfaces and emergent program technical issues required both additional program insight and additional funding in critical areas. WGS, with a firm fixed-price (FFP) contract that limits the cost impacts to the government, had problems with fasteners that delayed the delivery of the satellites by 15 months. A combined government and contractor team applied systems engineering to identify and prioritize mission-critical areas. This was an important step, especially in light of the fixed-price nature of the contract. The program encountered design, integration, and manufacturing problems due largely to the fact that the program was unable to benefit from continuing synergy on commercial SATCOM production lines. The Wideband Gapfiller System prime contractor had assumed a continued growth in the commercial SATCOM marketplace in its WGS FFP proposal. When the commercial SATCOM demand did not continue to grow, and in fact declined precipitously, the WGS prime contractor suffered substantial overruns on its FFP contract.

System Engineering Lessons

For the MILSATCOM Program to have been helped early by better systems engineering, each of the component elements would have needed to have been conceived in response to requirements from a consistent and coordinated set of users, advocated and funded by the same organization, and acquired, tested, and fielded in a coordinated way by an integrated program office. The current MILSATCOM Program was formed by combining preexisting satellite and terminal offices long after many important decisions had been made.

In both AEHF and WGS, the control and coordination processes that the MJPO has now put in place would have resulted in much earlier identification and resolution of issues and coordinated, joint problem resolution.

The AEHF program could have benefited greatly from earlier systems engineering that would have yielded a better understanding of the compatibility requirements with the legacy system and, in fact, understanding of the actual configuration of the sustained legacy Milstar system. A much more thorough analysis of the ability to accelerate AEHF in the face of immature technology would also have identified potential future problems and work-arounds. A more active man-

[12]Several important points are worth noting about the differences between the DOD MILSATCOM system and commercial SATCOM. As previously noted, MILSATCOM is a system of systems. While commercial SATCOM may consist of a family of incrementally improved satellites, these satellites do not and need not respond to the range of stressing missions accomplished by MILSATCOM. Commercial users are by and large less disparate and naturally have a more stable and narrower requirements set. Also, while commercial SATCOM cannot be considered "low-tech," it does not push the Technology Readiness Levels the way that MILSATCOM does. Finally, as a general rule, commercial SATCOM contracts are fixed-price production contracts, while DOD MILSATCOM contracts have been of the cost-plus type.

agement and engineering engagement with the National Security Agency (NSA) might have highlighted cryptographic issues sooner and allowed alternatives to be developed.

A better understanding of program interfaces and of the critical dependence of a program on Technology Readiness Levels might have led to different contracting decisions on the Wideband Gapfiller System. Active government program management and earlier joint systems engineering would have resulted in earlier and cheaper problem resolution.

While it is the mission of MJPO to deliver an integrated capability, many of the end users of the capability are not part of the Air Force Space Command (AFSPC), and their issues are not always well represented. The individual program managers do not and cannot control all aspects of key external requirements. While external users could always be better represented, the MILSATCOM SPO has made great strides in dealing with those users and understanding their issues. The MJPO has implemented a multipronged effort to provide a system-of-systems approach to the technical, business, and acquisition management of MILSATCOM products, and this approach is embodied in a wing-level systems engineering plan that addresses all issues at the system of systems level. Documented processes, carried out at the enterprise (wing) level, ensure that standards exist and are enforced among program (product) elements and also ensure that artifacts and information are generated to support cross-program integration.

By employing traditional systems engineering processes at the system of systems level, MJPO is able to reduce surprises and limit unintended consequences of individual program decisions and at the same time to gain an integrated view of gaps and overlaps among the product lines. The implementation and successful execution of such a systems-of-systems approach require leadership, experienced personnel, communication, and the willingness to make difficult trade-off decisions that affect individual program optimization.

C-5A Program

Description

The C-5A Program was characterized by an excellent pre-Milestone A and Milestone A-to-Milestone B process.[13] The Development Planning organization at Aeronautical Systems Division, Wright-Patterson Air Force Base, conducted operational mission analyses and contracted with industry (Lockheed Martin Corporation, The Boeing Company, and McDonnell Douglas Corporation) to conduct supporting conceptual design analyses. The Development Planning organization also conducted parallel conceptual design analyses of potential aircraft designs

[13]John M. Griffin, SES (Ret.), undated, *C-5A Galaxy Systems Engineering Case*, Wright-Patterson Air Force Base, Ohio: Center for Systems Engineering at the Air Force Institute of Technology. Available at http://www.afit.edu/cse/cases.cfm. Last accessed on June 18, 2007.

that could meet the evolving functional baseline. The user was an integral part of the process, as were laboratory scientists and technologists. The result of the entire process was a weapons systems specification that represented the functional baseline and was technologically feasible within the existing state of the art.

A SPO was established prior to the conduct of source selection; it was populated with domain experts in all of the functional areas. The SPO cadre included a number of people from the Development Planning Directorate who participated in the early development of the requirements and were knowledgeable regarding the genesis of the program.

In response to their perception of what their competitors would bid, Lockheed claimed an aggressively low value for its aircraft weight empty in its proposal. During the source selection, the Air Force convinced Lockheed to put the aircraft weight empty into its specification as a performance requirement equivalent to range and payload and other performance parameters. Then the Air Force put a financial penalty into the contract in the event that the contractor's aircraft weight exceeded the specification value, and Lockheed signed this contract. To complicate matters further, the contract type was a firm, fixed-price contract for both development and initial production under the new contract strategy that was being pursued by the Air Force, called Total Package Procurement. As a final constraint to the contract, a complementary financial penalty was included for each month that Lockheed missed the contract first flight date.

As the source selection came to a close, the three contractors were debriefed as to their strengths and weaknesses. The Air Force advised Lockheed that its assessment of the proposed aircraft showed the wing area to be deficient by some 400 square feet. Lockheed hastily redesigned the aircraft wing in just 4 days and resubmitted its proposal. The time available to redesign the wing was inadequate to conduct a systems engineering assessment of the new design; most of the technical parameters were updated as a ratio of the wing areas. This included the weight empty, which had been estimated with optimism for the original design and was now even more optimistic. In retrospect, this change contributed to a further disconnect of the cost estimate from the technical baseline. The net effect of all this activity during source selection was to excessively constrain costs, schedule, and performance to such tight and rigorous limits that there was no hope of being able to perform to the requirements of the contract.

As the design progressed to preliminary design review (PDR), it was clear to Lockheed that it could not meet the weight requirement. Lockheed proposed to the SPO that the weight requirement be removed from the contract, that the contractor would meet all of the performance requirements remaining in the specification, and that the engine contract would be increased by the $5 million needed to redesign the engine for an increase in engine thrust. The engine contractor was more than willing to redesign the engine for increased thrust, since its competitors had already seen the need to increase thrust to compensate for weight growth that was occurring in Boeing's commercial 747. The Air Force

System Program Office disapproved the request and, in fact, sent the contractor a cure notice.[14] Lockheed agreed to withdraw its offer and then directed its in-house engineering staff of designers to remove more weight from the structure of the aircraft. A competent system engineering assessment would have revealed that the resulting aircraft would be operationally unsuitable, having a limited load factor and reduced fatigue life. This was eventually recognized and accepted, and production was halted after 81 aircraft were completed.

Major reviews of the program and the design of the structure were conducted under the leadership of the Air Force. The combined government and industry team conducted an extensive systems engineering study of multiple options before selecting a new program approach. The new wing was redesigned to meet strength and life requirements. This was accomplished using basically the same design approach and adding 14,000 pounds to correct the problems. This new wing was retrofitted to the first 81 aircraft, and it was also included in the next buy of 59 C-5B aircraft. The Air Force spent over $2 billion more than the original budget, and it took 10 years longer to finish the production with fewer aircraft than originally planned. Lockheed petitioned the United States government through the Congress for a $250 million loan to avoid bankruptcy.

Systems Engineering Lessons

The C-5A case study makes the point that the pre-Milestone A and Milestone-A-to-Milestone B requirements process was efficient and effective. The C-5A Program should have succeeded based on the strength of the pre-Milestone A and Milestone A-to-Milestone B process. The point of failure on the C-5A Program occurred during source selection and was the result of poor application of the systems engineering process by both Lockheed and the Air Force during source selection after Milestone B. Because of the actions taken by the program participants, the program was doomed to failure from the first day of the contract. The lesson is that the systems engineering process in the pre-Milestone A and from the Milestone A-to-Milestone B time frame is an absolutely necessary function, but it is not sufficient. The process must be executed accurately and completely throughout the continuum of program acquisition.

Figure 2-3 shows the systems engineering graphic for the C-5A. It shows the systems engineering process in green for the requirements development phase and in red from contract go-ahead and on, and again turning to green for the retrofit and the C-5B, albeit at the expense of time and money.

---

[14]Before terminating a contract for default, the contracting officer may issue a written notice called a "cure notice." The notice allows the contractor 10 days to "cure" any defects. If the failure to perform is not cured within 10 days, the contracting officer may issue a notice of termination for default.

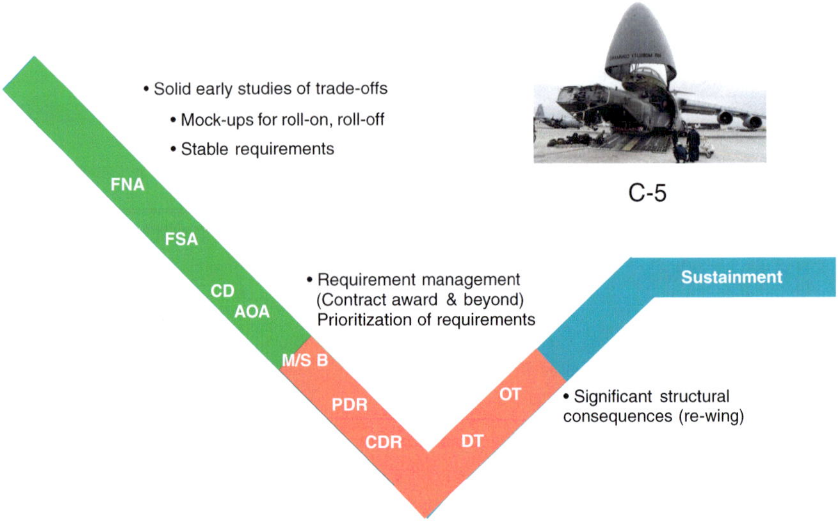

FIGURE 2-3 The C-5A systems engineering graphic showing an assessment of the application of the process. SOURCE: Contributed by committee member John Griffin.

B-2 Stealth Bomber Program

Description

As was the case with the C-5A discussed above, the B-2 case demonstrates a sound requirements process before Milestone A and successful systems engineering through Milestone B. The B-2 encountered problems just prior to PDR 2 when the design team completed an integrated analysis of the structure, flight controls, and aerodynamics. The design review showed that the aircraft had insufficient control power and structural rigidity to damp structural loads while providing the required maneuvering margins in the presence of turbulence.

The integrated design team, led by Northrop Grumman and including the major subcontractors, Boeing and Vought, and the Air Force customer, completed the redesign in 4 months and received approval for the new changes through the company and the Air Force management structure in 3 months. Notwithstanding its rapid systems engineering and design response to correct the problem, the subsystem design could not recover from the changes and caused a 6 month slip to the effective completion of critical design review (CDR). In this case, the maturity of the design and confluence of three technical disciplines in a new concept led to the realization that a redesign was necessary. The design team quickly developed a new approach, and the management team responded with new plans. However, the subsystems designers and the subcontractors did not have sufficient time to recover their previous work and were late to CDR. This was not a failure in requirements, but rather a recovery of the design and a lag in recovery by one segment of the team.

Systems Engineering Lessons

The B-2 case study illustrates that even when the early systems engineering process is done well, the acquisition process is fraught with peril because of the unknowns and complications that arise in any program.

However, from the case studies and from listening to the testimony of the briefers, it is clear that program offices and industry teams staffed with domain experts equipped to handle technical and programmatic difficulties are best suited to respond quickly and effectively to the problems when they arise. Managers who are inexperienced in handling problems such as occurred in this case may be unable to prevent the full collapse of a program. The acquisition process is by nature a complicated and delicate one that can easily be driven to instability. This point is underscored and somewhat amplified by comparing the C-5A participants' actions and those taken by the B-2 team. The C-5A Program certainly had domain experts, as did the B-2. However, the C-5A Program was complicated by the extreme controversy of the program, the fixed-price Total Package Procurement contract approach, the contractor's aggressive claims, and the Air Force's reluctance to re-open the contract after it was signed, all of which combined in a complicated dynamic and contributed to the less-than-domain-expert decisions. In the B-2 case, the program proceeded with the agreement of all parties, notwithstanding that the Northrop Grumman program management stated concern that the schedule of CDR was at risk. This risk was eventually realized, but it was not a requirements issue. Rather, it was a technical issue that surfaced as a result of the need for the aircraft redesign to meet the requirements on contract from the start of Milestone B.

Figure 2-4 presents the systems engineering graphic for the B-2, showing that it turns red at PDR and then recovers to green at CDR, albeit a year late and nearly $1 billion over the estimate at completion.

SHARED FINDINGS AND LESSONS LEARNED AMONG CASES

Although each of the case studies summarized above revealed some unique findings, several key findings and lessons learned are shared among the programs. These are summarized here:

- There is a need for an appropriate level of SE talent and leadership early in the program, with clear lines of accountability and authority. Senior SE personnel should be experienced in the product(s) domain, with strong skills in architecture development, requirement management, analysis,

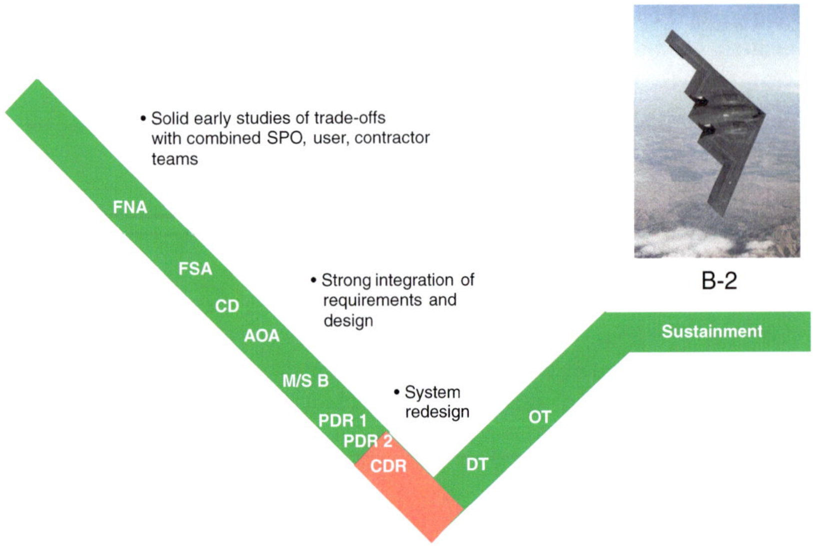

FIGURE 2-4 The B-2 systems engineering graphic showing an assessment of the application of the process. SOURCE: Contributed by committee member John Griffin.

modeling and simulation, affordability analysis, and specialty engineering disciplines (e.g., reliability, maintainability, survivability, system security, and technology maturity management).

- There is a need to establish and nurture a collaborative user/acquirer/industry team pre-Milestone A to perform system trade-offs and manage overall system complexity. Today, there are often significant disconnects in the hand-offs between users, acquirers, requirements developers, industry, and others. Some of the "best practices" include structured collaboration among these members.
- One must clearly establish a complete and stable set of system-level requirements and products at Milestone A. While requirements creep is a real problem that must be addressed, some degree of requirements flexibility is also necessary as lessons involving feasibility and practicality are learned and insights are gained as technology is matured and the development subsequently proceeds. Certainly control is necessary, but not an absolute freeze. Also, planning ahead for most likely change possibilities through architectural choices should be encouraged, but deliberately managed, a concept encouraged herein. A typical program execution team has a program manager (PM)-level SE integration team (SEIT), with

responsibility, authority, and accountability to perform the SE functions (including analysis, modeling and simulation, architecture development, requirements management, and so on). Some of the "program discipline" needs to be in pre-Milestone A management.

- It is necessary to manage the maturity of technologies prior to Milestone B and to avoid reliance on immature technologies. Technology maturity and risk mitigation plans should be carefully managed as an integral part of program plans.

The above represent lessons learned as a result of problems that arose or successes that were achieved on past or current programs. They are applicable in general. It is crucial for programs currently being formulated or beginning the acquisition process, TSAT and Space Radar being cases in point, that these lessons be applied early. It is incumbent on senior operational and acquisition leadership to enforce the discipline implied by these findings and shared lessons.

3

Systems Engineering Workforce

INTRODUCTION

As illustrated by several of the case histories described in Chapter 2 (particularly those of the F-16, the fighter jet engine program, and the B-2), the presence of experienced, domain-knowledgeable systems engineers on the development team—on both the government and the industry sides—is a critical factor in the success of any Air Force acquisition program. However, in recent years the depth of systems engineering (SE) talent in the Air Force has declined owing to policies within the Department of Defense (DOD) that shifted the oversight of SE functions increasingly to outside contractors, as well as to the decline of in-house development planning capabilities in the Air Force (AF). The result is that there are no longer enough experienced systems engineers to fill the positions in programs that need them, particularly within the government. As acquisition programs continue to evolve from individual systems to systems of systems, this shortage will only become more acute.

For the Air Force to be a "smart buyer" of systems and systems modification programs, its personnel must be well trained to supervise and critically evaluate progress in the various programs. The Air Force needs personnel qualified to anticipate problems and respond intelligently to them. The Air Force cannot outsource its technical and program management experience and intellect and still expect to acquire new systems that are both effective and affordable.

This chapter discusses the U.S. SE workforce in terms of the production of systems engineers by U.S. universities, industry, and the Air Force. The approaches taken by industry to train systems engineers are described and, where there are specific areas of emphasis, these are noted. The duty assignments of Air Force systems-engineering-trained officers and civilians are described. For the

U.S. Air Force Academy (USAFA) and Air Force Institute of Technology (AFIT), the chapter presents data on the number of systems engineering graduates and their follow-on assignments. The chapter also addresses the numbers of officers trained in systems engineering that the Air Force expects to have in the future. This is particularly important given the manpower drawdown that the Air Force is going through as a result of Program Budget Decision (PBD) 720.[1]

As best the committee can determine, the Air Force does not have systems engineers assigned between Milestones A and B; hence, the committee concludes that none are assigned in the pre-Milestone A period. Furthermore, as discussed later in this chapter, the personnel/manpower "accounting" system that the Air Force uses does not enable the easy tracking of personnel who are performing SE functions or jobs that require them. Hence it is nearly impossible to assess supply and demand for systems engineers.

PRODUCTION OF SYSTEMS ENGINEERS BY U.S. UNIVERSITIES

Figure 3-1 shows that the output of systems engineering degrees in U.S. universities has increased slowly over the past decade.

This conclusion is supported by data cited in a forthcoming report by the International Council on Systems Engineering (INCOSE),[2] which is developing a reference curriculum for systems engineering. Engineering schools such as the Georgia Institute of Technology (Georgia Tech), Massachusetts Institute of Technology (MIT), and Stevens Institute of Technology are introducing new professional and executive master's degree programs in systems engineering and systems management based on this INCOSE reference model. The curriculum places a strong emphasis on domain expertise (e.g., electrical engineering, mechanical engineering) at the undergraduate level.

Figure 3-1 includes data for systems-engineering-centric programs only. It does not include domain-centric systems engineering programs. For example, universities such as Stanford University, Georgia Tech, and the California Institute of Technology have exceptional programs in aerospace engineering, electrical engineering, and industrial engineering that include aspects of systems engineering.[3]

[1]Program Budget Decision 720, entitled "Air Force Transformation Flight Plan," was issued on December 28, 2005, by the Under Secretary of Defense (Comptroller). In it, the Defense Comptroller directed reductions in Air Force manpower from 2007 to 2011 totaling over 40,000 people, including active, Air National Guard, and Air Force Reserve civilian, officer, and enlisted personnel. Manpower reductions in specific career fields were not specified in the PBD, but it is expected that the scientist, engineer, and acquisition manager career fields will experience significant reductions as the PBD 720 reductions are allocated.

[2]R. Jain and D. Verma, 2007, *Proposing a Framework for a Reference Curriculum for a Graduate Program in Systems Engineering*, Hoboken, N.J.: International Council on Systems Engineering.

[3]W. Fabrycky and E. McCrae, 2005, *Systems Engineering Degree Programs in the United States,* Hoboken, N.J.: International Council on Systems Engineering.

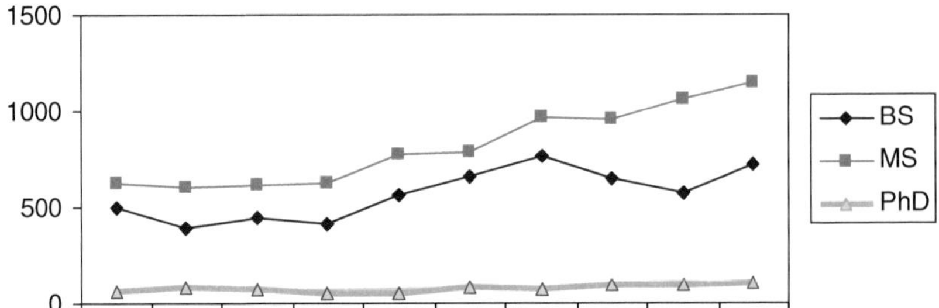

	1997	1998	1999	2000	2001	2002	2003	2004	2005	2006
BS	503	393	442	413	568	659	763	649	570	723
MS	632	609	614	626	773	782	970	953	1065	1150
PhD	67	89	75	54	52	81	75	92	94	104

FIGURE 3-1 Systems engineering degrees awarded in the United States in the past decade. SOURCE: Based on data gathered in Engineers Joint Council, Engineering Manpower Commission, and American Association of Engineering Societies, 2006, E*ngineering and Technology Degrees*, New York: Engineering Manpower Commission.

Thus, Figure 3-1 does not present a complete picture of U.S. university production of engineers that have an exposure to systems thinking.

PRODUCTION OF SYSTEMS ENGINEERS BY U.S. INDUSTRY

Industry has clearly recognized the need for SE-trained personnel. In fact, it has invested significantly in training programs that produce hundreds or even thousands of company-trained systems engineers per year. The committee interviewed representatives from four major U.S. aerospace companies to better understand industry approaches used to develop and train systems engineers. To protect the proprietary nature of any of the approaches being used, the companies themselves are not identified in the report. Some of these companies have emphasized systems engineering training for less than a decade, while others have been involved in it for as long as 30 years. The discussion below summarizes common themes that emerged from the interviews.

- *Training, not just education, is crucial.* All the companies agree that a person learns to be a systems engineer by on-the-job-training (OJT)—by practicing the trade. While tools that facilitate the management of a program can be taught and learned, the essence of being a good systems engineer depends on applying all knowledge, including functional and domain knowledge, along with the tools, at the right places in any given program. The skill is sharpened through experience, and both success and failure are good teachers.

- *All the companies agree that mentoring is essential.* This is especially true when the loss of experienced personnel occurs and the next level of personnel must be developed as quickly as possible. At the same time, most of the companies are aggressively documenting those practices and processes that are time-tested and essential so that these processes can be available to those who are learning, and not lost when the key personnel retire.
- *Subject matter expertise and/or domain knowledge are more important than is a knowledge of tools.* The foundation for a good systems engineer is his or her academic training in a technical area (e.g., aeronautical engineering, electrical engineering, or software engineering), augmented by OJT. The tools that are taught and acquired are a means to an end—necessary, but by no means sufficient. A person who is trained only in the tools of systems engineering is not a systems engineer.
- *Both internal and external training are valuable; the most successful training approach is usually a hybrid.* In general, the companies find that schools provide useful training but rarely provide the kind of insight that a tailored in-house program does. They also find that, for the most part, the cost of in-house training is on a par with the cost of university training. Of interest is the fact that some universities (e.g., the University of Southern California [USC]) have created positions called Professor of Practice. These nontenured positions are specifically designed to enable the hiring of practitioners of a given skill or craft to augment the regular faculty. At USC, professors of practice are hired in systems engineering, among other areas.
- *Certification by and participation in INCOSE are considered essential.* All the companies require certification (acquired through the right training and experience), and all participate in and support INCOSE.
- *Investment in SE training is necessary whether or not the return on investment can be directly estimated.* Some of the companies have been able to quantify their return on investment (ROI) for the training—they estimate or calculate the benefit, given the cost. All of them say, though, that they cannot compete without the training, even if they cannot directly estimate its ROI.
- *A systems engineering culture is essential.* All the companies agree that there must be a culture of systems engineering and that it must pervade every program, no matter how large or small. If the small programs are neglected, this can lead to problems and failure, which cost the company time and money to correct. The prevailing view is that systems engineering is not a phrase, a bumper sticker, an organization, or a job code—systems engineering is a discipline. It is not something that one can have

a nodding acquaintance with; nor is it something that one can just be familiar with. It is something one has to own and believe in.

- *Systems engineering organizations vary.* Some of the companies interviewed have separate systems engineering groups or departments. In contrast, one company disbanded its systems-engineering-specific organization and dispersed the professionals to all functional levels. The reason given was that having a separate organization led to perceptions among Integrated Product Team (IPT) leaders that systems engineering responsibilities are handled by the "systems engineering group." This caused these leaders to neglect the critical role and responsibility that they themselves had for implementing the systems engineering development environment.
- *The "trigger" for a company's emphasis on systems engineering is usually failing programs.* Faced with several troubled programs, an analysis typically revealed that there was a fundamental lack of systems engineering in all of them or, if it was present at all, it was not being applied correctly. It was also often observed that the personnel who claimed to be systems engineers were insufficiently trained, including the managers who claimed to be systems engineers or to have training in it.

THE ROLE OF FEDERALLY FUNDED RESEARCH AND DEVELOPMENT CENTERS

Systems Engineering FFRDCs

The Aerospace Corporation and the MITRE Corporation are Air Force systems engineering federally funded research and development centers (FFRDCs). The FFRDCs provide independent, objective, credible support and work to the Air Force customers for whom they work.

Each of them is allocated a total number of staff technical equivalents (STEs)—referred to as a ceiling—as a limit to which they can be funded. While not all of the work that they do is for the Air Force, the lion's share of Aerospace's is for the Air Force, while less than half of MITRE's is assigned to the Air Force. Specifically, for Aerospace, approximately 89 percent of its total ceiling was allocated to the Air Force for fiscal year (FY) 2006, and 88 percent for FY 2007. For MITRE, approximately 49 percent of its total ceiling for FY 2006 was allocated to the Air Force Electronic Systems Center (ESC) and non-ESC Air Force organizations, and approximately 46 percent of its total ceiling for FY 2007.[4]

[4]Data on both the Aerospace Corporation and the MITRE Corporation were provided in a private communication on April 4, 2007, between the committee and Michael Kratz, Chief of Acquisition Workforce Policy and Resources at SAF/AQX. Note that these figures were based on in internal review and do not include National Intelligence Program exclusions nor FY 2007 AF Military Intelligence Program exclusions. Also, in FY 2006, STE was placed on contracts with $780.5 million allocated to Aerospace and $250.9 million for AF allocation to MITRE.

The committee was not able to obtain a breakout showing what functions these engineers are performing for their Air Force customers. Probably not all are performing systems engineering functions; however, these numbers represent upper bounds.

Studies and Analysis and "Technology Transition" FFRDCs

Studies and analysis FFRDCs such as the RAND Corporation and the Institute for Defense Analyses (IDA) have played and can play an important role, particularly in the pre-Milestone A period. In the acquisition process, analysis needs to be done early (and continuously) to help frame the boundaries of requirements and system performance and to contribute to important knowledge and understanding at the intersection of operational needs analysis and technical solution analysis. These activities have been and should remain complementary to any Air Force requirements organizations. "Technology transition" FFRDCs, such as MIT's Lincoln Laboratory and Carnegie Mellon University's Software Engineering Institute (SEI), can contribute importantly as they focus on—and transition—best practices related to systems and software engineering. The committee saw evidence of these capabilities during the briefings that it received.

SYSTEMS ENGINEERING TRAINING AND EDUCATION WITHIN THE AIR FORCE

There are two Air Force institutions that provide formal systems engineering training—the AFIT and the USAFA. The AFIT program and the intense interest in systems engineering by Secretary of the Air Force James G. Roche was the stimulus for creating the USAFA program. While the USAFA program is at the undergraduate level only, it does teach the students principles of systems engineering, and they have to complete a senior project that is multidisciplinary and allows them to apply the aspects and elements of systems engineering at some level.

The Genesis of the Air Force Center for Systems Engineering

In the spring of 2002, while meeting with the commander of the Air Force Materiel Command (AFMC) and later with the commandant of the AFIT, Secretary of the Air Force Roche directed that an organization be created to help strengthen the Air Force's systems engineering capabilities. Further, he directed that this organization be led by a general officer or civilian equivalent and be located at AFIT at Wright-Patterson Air Force Base near Dayton, Ohio.

Following up on that direction in the fall of 2002, AFMC conducted a systems engineering forum bringing together 54 of the leading systems engineering experts in the country. The forum identified key gaps and shortfalls in the defense industry systems engineering community and provided recommendations to address those gaps and shortfalls. Additionally, the forum members discussed possible roles for an Air Force organization that could address the gaps and options and a structure for that organization. By the end of 2002, the commanders of the AFMC, the Air Force Space Command (AFSPC), and the Air Education and Training Command (AETC) decided that a new organization would be formed and that it would belong to AETC located at AFIT. Its director would be a member of the Senior Executive Service and would report directly to the AFIT commandant. The commanders also pledged to find positions from all three commands to staff the organization.

Thus, the Air Force Center for Systems Engineering (CSE) was born in early 2003. The center director was the equivalent of a dean at AFIT, and the center had its own governing council. The focus for the center's activities grew out of the recommendations of the systems engineering forum and included education, training, collaboration, and advocacy. As the center matured in the following months, it focused on two goals:

- *To influence and institutionalize the systems engineering process.* This goal includes an in-house rotational program for development of new systems engineers, consultation with other organizations, and the development of systems engineering tools, processes, and practices in collaboration with organizations such as INCOSE.
- *To educate the workforce.* This goal includes the development of systems engineering case studies; graduate programs; seminars, workshops, and short courses on systems engineering and architecture; and initiatives to provide accessibility to these programs at key locations throughout the Air Force.

The Air Force CSE has delivered on the goals outlined above and has published comprehensive case studies on programs that include the C-5, F-111, Hubble Space Telescope, Theater Battle Management Core System, B-2, and the Joint Air-to-Surface Standoff Missile (JASSM). It has published several SE guides and is an active participant in numerous systems engineering venues and initiatives. Additionally, AFIT has produced more than 200 graduates of its master's and certificate programs in systems engineering and architecture since the center was formed in 2002. The center and AFIT collaborate with numerous universities on curricula, joint graduate capstone projects, delivery of the Graduate Certificate in Systems Engineering, and many short courses.[5] Recently AFIT expanded its utilization of distance learning technology to make its courses available to more individuals and organizations across the nation.

---

[5]A current list of AFIT graduate capstone projects can be found at http://www.usafa.af.mil/df/dfsem/Capstones.cfm?catname=dean%20of%20faculty. Last accessed on April 27, 2007.

TABLE 3-1 Air Force Institute of Technology (AFIT) Intermediate Development Education (IDE) Students, 2004-2008

Graduation Year	Total IDE Class[a]	IDE SE Students[a]
2004	80	21
2005	140	37
2006	220	35
2007	80[b]	12[b]
2008	43[c]	21[c]

[a]The selection of Air Force officers for IDE at AFIT is a function of the number of officers designated by promotion boards to receive IDE, and the subsequent selection of officers to go to particular schools from those listed for a given year. As the number of officers in a year group goes down, the number being designated for IDE by any promotion board will decrease as well.

[b]Slated to graduate.

[c]Inbound. Also, starting in 2007, the program had 4 versus the 12 options that were previously available.

SOURCE: Air Force Institute of Technology.

In addition, one option for Air Force intermediate development education (IDE) is to attend AFIT and obtain a master's degree in addition to professional military education. One of the master's degrees available is that of systems engineering. The first class graduated from this program in 2004. Table 3-1 shows the total numbers in these IDE classes (actual for years 2004 through 2006, slated to graduate in 2007, and planned for 2008) and of those how many received a master's degree in systems engineering or plan to do so.

U.S. Air Force Academy Training in Systems Engineering and Systems Engineering Management

The Air Force Academy has two systems engineering majors: systems engineering and systems engineering management (SEM) (the latter is not accredited). Cadets in both majors get experience applying their specialties by teaming up with engineering domain-specific cadets in one of nine defined Capstone Design projects in the following departments: Aeronautical Engineering, Astronautical Engineering, Civil Engineering, Computer Science, Electrical Engineering, Engineering Mechanics, and Operations Research.

The first year that cadets graduated with degrees in these majors was 2006. The numbers of graduates for that year and those projected to graduate with these majors in 2007, 2008, and 2009 are shown in Table 3-2.

Those who graduated in 2006 were assigned to the career fields shown in Table 3-3. Also shown are the assignments for those cadets who were expected to graduate with these majors in 2007. Both are summarized in Table 3-4.

TABLE 3-2 U.S. Air Force Academy Graduates with Majors in Systems Engineering and Systems Engineering Management, 2006-2009

Year	Systems Engineering[a]	Systems Engineering Management[b]
2006	32 (graduated)	68 (graduated)
2007	43 (projected)	91 (projected)
2008	51 (projected)	99 (projected)
2009	42 (projected)	67 (projected)

[a]Up for initial ABET, Inc., accreditation in 2008.
[b]No plans to accredit.
SOURCE: U.S. Air Force Academy.

TABLE 3-3 Career Fields to Which U.S. Air Force Academy Graduates in 2006 and 2007 in Systems Engineering (SE) and Systems Engineering Management (SEM) Were Assigned

	CL2006		CL2007[a]	
Class	SE	SEM	SE	SEM
32E1G Civil Engineer	1			
33S1 Communications and Information	2	3		3
61S1A Scientist	1	3		
62xxx Development Engineer	2		9	
63xxx Acquisition Manager	1	9		10
92M1 Medical Student	1			
92T0 Pilot Trainee	19	45	29	53
92T1 Navigator Trainee	1	1	3	1
Army	1			
13M1 Air Field Operations	1			2
13S1 Space and Missile		1		4
41A1 Health Services Administrator		1		
64P1 Contracting		1		5
65F1 Financial Management		2		5
65W1 Cost Analysis		1		
21A1 Aircraft Maintenance		1	2	4
21R1 Logistics Readiness				1
14N1 Intelligence				3
Total	32[b]	68	43	91

[a]Class of 2007 assignments are projected.
[b]Assigned career fields for two 2006 SE graduates were unspecified.
SOURCE: U.S. Air Force Academy.

TABLE 3-4 Summary of Systems Engineering (SE) and Systems Engineering Management (SEM) Assignments (Classes of 2006 and 2007)

Assignment	Total Assignments (Percent)
Rated	65
Operations (Air Field Operations/Space and Missile/Maintenance)	12
Technology (Scientist/Engineer/Communication/Information)	10
Contract/Finance/Cost Analysis	6
Acquisition	4
Other (Intelligence, Logistics, Health)	3

SOURCE: U.S. Air Force Academy.

The goals of a new program to enhance engineering education at the Air Force Academy are to (1) encourage underclass cadets to major in ABET-accredited engineering disciplines, (2) motivate upperclass cadets to pursue Air Force careers in engineering, and (3) support improvement and expansion of the USAFA systems engineering program.[6] The proposed approach includes fall and spring semester onsite lecture series and mentoring involving junior active-duty Air Force engineers.

CURRENT INVENTORY OF AIR FORCE OFFICERS ASSIGNED AND TRAINED IN THE SCIENTIST, ENGINEER, AND ACQUISITION MANAGER CAREER FIELDS

Systems engineering expertise derives from initial academic training to obtain domain expertise, postgraduate training to deepen domain experience, postgraduate training to learn how to use systems engineering management tools, and hands-on experience in program development and management. The formal training for systems engineers and the overt recognition of the importance of SE as an Air Force competency is a relatively recent occurrence. Perhaps that, coupled with the fact that there is no undergraduate degree in SE, is why there is not yet a classification code for systems engineers in the Air Force.

Figures 3-2, 3-3, and 3-4 show, respectively, the numbers of officers in the 61S (scientist), 62E (engineer), and 63A (acquisition manager) career fields by years of service and grade. The number of officers in the 61, 62, and 63 career fields diminishes rapidly with increasing grade and years of service.

The numbers of scientific and engineering Air Force officers are shown in Figure 3-5 for five areas of engineering (aerospace, astronautical, computer,

[6]In 2006, Paul Kaminski, a graduate of the Air Force Academy, made a gift to the Association of Graduates to support and improve engineering education and enrollment at the Academy.

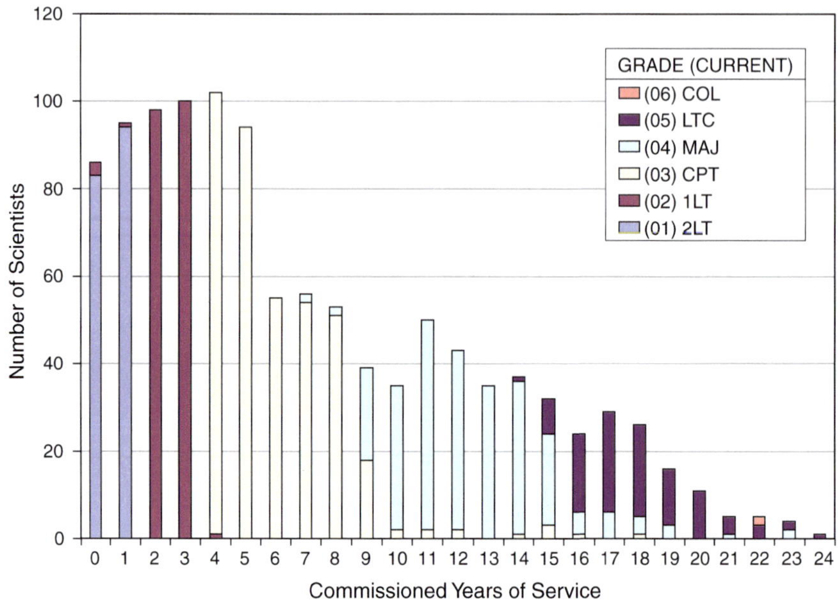

FIGURE 3-2 Number of scientists (61S) by grade and years of service. SOURCE: Air Force Personnel Center, Interactive Demographic Analysis System (IDEAS), http://w11.afpc.randolph.af.mil/vbin/broker8.exe?_program=ideas.IDEAS_default.sas&_service=vpool1&_debug=0 (as of March 2007).

electrical, and mechanical engineering), the level of degree (BS, MS, and PhD), and where the officers were assigned. Since there is no tracking of officers who have any education in SE or experience applying it, it is not possible to make that distinction from these data. Note the very small numbers of engineers in the 63A career field, where program managers would be found.

The only way that an officer with academic SE training can be found in the Air Force personnel database is by searching for the degree type. The shortcomings of this process are that there is not a consistent description of degrees, and it is very time-consuming. Further, this type of search does not reveal if a person so trained has had any successful hands-on experience in the application of systems engineering principles.

AIR FORCE CIVILIAN SYSTEMS ENGINEERING POSITIONS

Most of the engineering positions in the Air Force are in the materiel and space commands (AFMC and AFSPC, respectively), in which the bulk of pro-

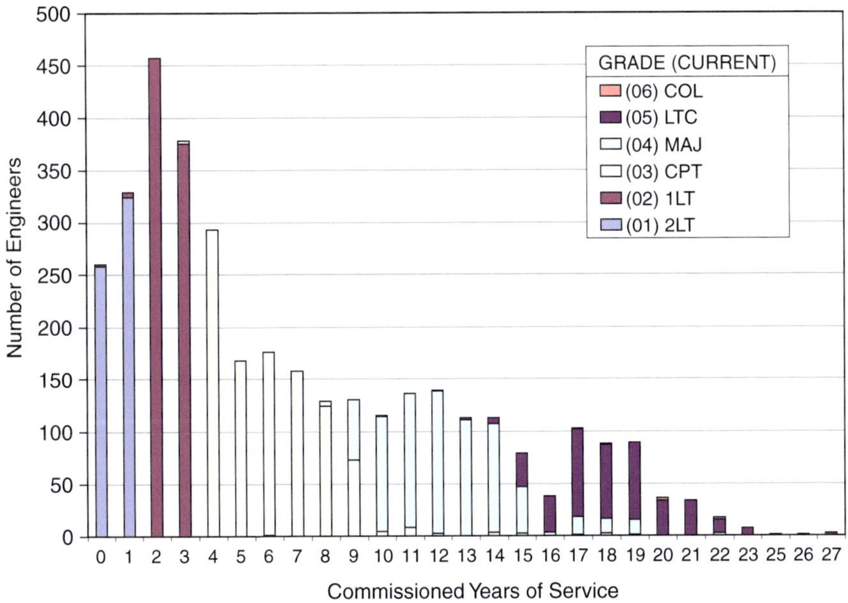

FIGURE 3-3 Number of engineers (62E) by grade and years of service. SOURCE: Air Force Personnel Center, Interactive Demographic Analysis System (IDEAS), http://w11.afpc.randolph.af.mil/vbin/broker8.exe?_program=ideas.IDEAS_default.sas&_service=vpool1&_debug=0 (as of March 2007).

gram acquisition and development takes place. The committee found it difficult to obtain any data on SE positions for AFSPC, but some data exist for AFMC, as discussed below.

AFMC Instruction (AFMCI) 62-202 (*AFMC Core Criteria for Critical Engineering Positions*) delineates four position codes for these critical engineering positions: Lead Engineer (806), Chief Engineer (805), Director of Engineering (807), and Technical Director (356). These are leadership positions, are not at the Senior Executive Service level, but do apply to civilian positions at the GS-15 level and equivalent and below, to military positions at the rank of colonel and below, and to equivalent contractor positions. There are undoubtedly other engineers, occupying positions lower than these critical engineering positions, who are performing some systems engineering tasks in nonleadership roles; however, it would be extremely difficult to break them out because the Air Force has no position identifier for "systems engineering." Thus, this discussion is limited to positions of systems engineering leadership.

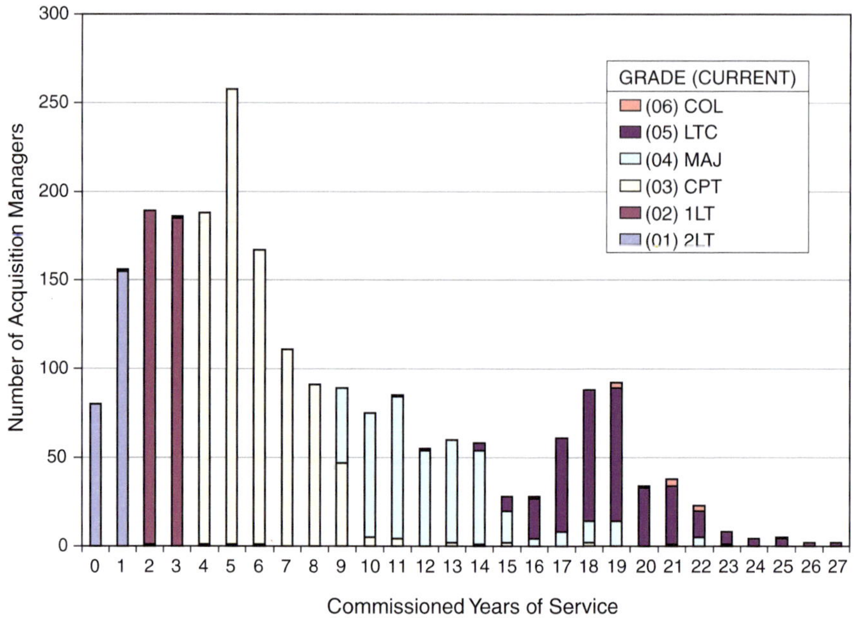

FIGURE 3-4 Number of acquisition managers (63A) by grade and years of service. SOURCE: Air Force Personnel Center, Interactive Demographic Analysis System (IDEAS), http://w11.afpc.randolph.af.mil/vbin/broker8.exe?_program=ideas.IDEAS_default.sas&_service=vpool1&_debug=0 (as of March 2007).

The definitions in the AFMCI for each position type are as follows:

- *Lead engineer*. Engineer responsible for a single end item, or family of end items; has operational safety, suitability and effectiveness (OSS&E) responsibility; responsible for all end item/commodity technical activities, including engineering configuration changes.
- *Chief engineer*. Senior engineer/technical authority for a weapon system or equivalent product; has OSS&E responsibility.
- *Director of engineering*. Senior engineer/technical authority responsible for multiple chief or lead engineering positions; ensures programs under their purview are addressing OSS&E; ensures chief and lead engineers assigned to systems/end items within their organization are executing their responsibilities appropriately; fulfills chief engineer responsibilities for systems/end items without an assigned chief engineer.
- *Technical director*. Senior engineer; technical specialty position for engineering; provides expertise on technical aspects supporting directorate or wing operation and processes; has various levels of OSS&E responsibility.

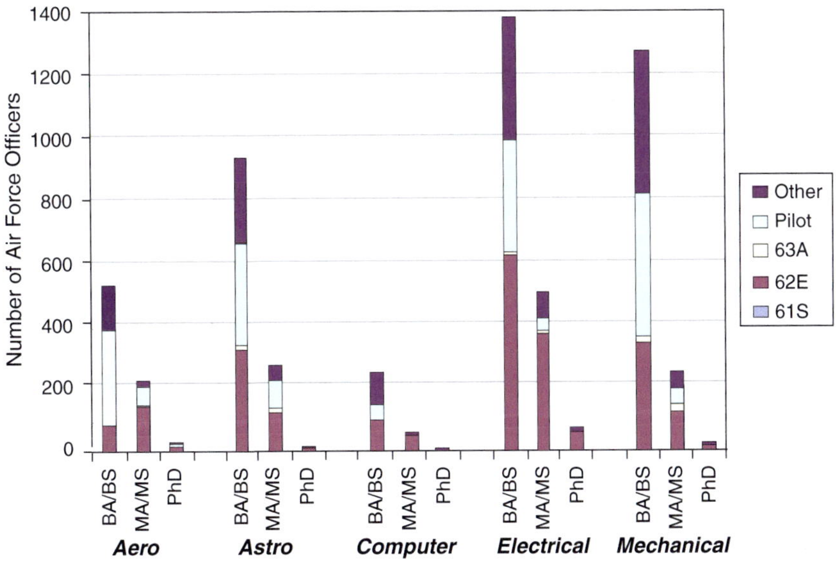

FIGURE 3-5 Air Force officers with engineering degrees: breakdown by field of specialization and career track. SOURCE: Air Force Personnel Center, Interactive Demographic Analysis System (IDEAS), http://w11.afpc.randolph.af.mil/vbin/broker8.exe?_program=ideas.IDEAS_default.sas&_service=vpool1&_debug=0 (as of March 2007).

Lead engineer positions are most likely domain-centric; thus individuals in these positions would not necessarily be working at the systems level and performing systems engineering tasks in a leadership role as discussed here. Individuals in the chief engineer, director of engineering, and technical director positions probably are doing work at the systems level and thus are performing systems engineering leadership tasks.

As of March 2007, AFMC had 231 of these critical engineering positions, with 180 at product centers and the remaining 51 at logistics centers.[7] All are organic, with the exception of 43 at ESC, which are staffed with individuals from MITRE, an FFRDC. Not all are civilian, but most are. Unfortunately, the data from AFMC do not break out the military positions.

Breaking out these positions by code puts 84 in the lead engineer category, 88 in the chief engineer category, 36 in the director of engineering category, and 23 in the technical director category.

[7]Personal communication between committee member Mark K. Wilson and Dominick Tucillo, Air Force Materiel Command Engineering Directorate.

Not counting those in the lead engineer category and adding one each for the AFMC centers for their director of engineering SES positions (6 when SMC is included), two positions for the AFMC Engineering Directorate (AFMC/EN) (director and technical director), and six positions at the Air Force CSE (including its director), the total number of "systems engineering leadership" positions is 161. Therefore, it is estimated that approximately 160 of the engineering positions in the Air Force (if filled) have incumbents who are performing systems-engineering-related work in a leadership role. Note also that AFMC Instruction 62-202 establishes the criteria to be used in the selection process for individuals in these critical engineering positions.

EFFECTS OF PROGRAM BUDGET DECISION 720

The Air Force has made a concerted effort to access, retain, and shape the military scientist and engineer (S&E) career fields. Following low retention years in the late 1990s and early 2000s, the Air Force offered its scientists and engineers retention bonuses and increased the S&E accession levels well above what was required on a steady-state basis to maintain a healthy force. These efforts had a dramatic effect between 2001 and 2005, increasing the manning of these career fields as shown in Table 3-5.

However, after PBD 720 is fully implemented, the numbers of officers accessed into Air Force 61S and 62E career fields are expected to be smaller. As a result of PBD 720, the Air Force has taken a hard, detailed look at each of its officer career fields. The S&E share of the authorization cuts are shown in Table 3-6.

To shape the inventory to match these reductions while taking into account the future health of the force, the Air Force has used a steady-state sustainment methodology. In the steady state, accession levels are set to provide 100 percent manning over the course of 30 years—constant accessions against a constant retention rate, calculated distinctly for each career field. During the force-reduction years, the Air Force planned to meet required end strength by accessing each specialty at the sustainment level and managing losses through primarily voluntary losses. The S&E community is currently planned to access at 91 percent of sustainment through FY 2009, after which it will return to the sustainment level. Table 3-7 shows for the 61S (scientists) and 62E (engineers) career fields the actual accessions for FY 2001 through FY 2006 and the planned accessions for years FY 2007 through FY 2009. Also shown are the annual sustainment targets for each career field, goals that will sustain a healthy force for the future while allowing the Air Force to meet end-strength targets during this period of significant force drawdown.

TABLE 3-5 Increase in the Number of Air Force Scientists and Engineers from 2001 to 2005

Career Field AFSC	2001 Inventory	2001 Authorizations	2001 Manning	2005 Inventory	2005 Authorizations	2005 Manning
61S (Scientist)	753	923	82%	889	899	99%
62E (Engineer)	2,072	3,045	68%	2,614	2,702	97%

SOURCE: Headquarters U.S. Air Force /A1.

TABLE 3-6 Current and Projected Cuts in Air Force Science and Engineering Personnel Resulting from Program Budget Decision 720

	61S (Scientist)			62E (Engineer)		
FY	Permanent Party Authorizations	% Change	Cumulative % Change	Permanent Party Authorizations	% Change	Cumulative % Change
2006	877			2,654		
2007	796	–9.20	–9.20	2,405	–9.40	–9.40
2008	791	–0.60	–9.80	2,411	0.20	–9.20
2009	768	–2.90	–12.40	2,384	–1.10	–10.20

SOURCE: Headquarters U.S. Air Force /A1.

TABLE 3-7 61S and 62E Actual and Planned Accessions and Sustainment Targets

Career Field	Sustainment Target[a]	Actual Accessions 2001	2002	2003	2004	2005	2006	Planned Accessions 2007	2008	2009
61S (Scientist)	70	69	94	128	94	91	90	97	64	64
62E (Engineer)	297	155	196	326	371	386	312	346	272	272

[a]Number of qualified accessions required to meet future force authorizations (using FY 2011 authorizations from the FY 2007 President's Budget end strength; authorizations remain the same from FY 2009 through FY 2011).

SOURCE: Headquarters U.S. Air Force /A1.

CONGRESSIONAL ACTIONS TO CUT DOD ACQUISITION WORKFORCE

The PBD 720 workforce cuts that the Air Force is taking are not the first significant reductions in the acquisition workforce. In fact, Congress participated with direction in the FY 1996, FY 1997, FY 1998, and FY 1999 Defense Authorization Acts to reduce the DOD acquisition workforce. The reductions mandated by Congress in FY 1996 put the Air Force's acquisition workforce on a precipitous path.

The Federal Acquisition Streamlining Act (FASA) II (called the Federal Acquisition Reform Act [FARA]) was passed during the first session of the 104th Congress. It built on the earlier FASA legislation and was included in the FY 1996 DOD Authorization Act (P.L. 104-106). The newest reform provisions sought (1) to simplify procedures for procuring commercial products and services, and at the same time to preserve the concept of full and open competition; (2) to reduce barriers to acquiring commercial products by eliminating the requirement for certified cost and pricing data for commercial products; and (3) to streamline the bid protest process by providing for all bid protests to be adjudicated by the General Accounting Office (GAO; now the Government Accountability Office). To reflect the projected efficiencies of acquisition reform and the broader personnel reductions occurring at DOD, FASA directed DOD to reduce its acquisition workforce by 15,000 personnel during FY 1996 and to report to Congress on how to implement an overall 25 percent reduction during the next 5 years (from October 1, 1995).

In the FY 1997 Defense Authorization Act, Congress directed an acquisition workforce reduction of an additional 15,000 in Section 902 of the act:

> SEC. 902. ADDITIONAL REQUIRED REDUCTION IN DEFENSE ACQUISITION WORKFORCE.
>
> (a) Additional Reductions for Fiscal Year 1997.—Section 906(d) of the National Defense Authorization Act for Fiscal Year 1996 (Public Law 104–106; 110 Stat. 405) is amended in paragraph (1) by striking out "positions during fiscal year 1996" and all that follows and inserting in lieu thereof "so that—"(A) the total number of defense acquisition personnel as of October 1, 1996, is less than the baseline number by at least 15,000; and "(B) the total number of defense acquisition personnel as of October 1, 1997, is less than the baseline number by at least 30,000.".
>
> (b) Baseline Number.—Such section is further amended by adding at the end the following new paragraph: "(3) For purposes of this subsection, the term 'baseline number' means the total number of defense acquisition personnel as of October 1, 1995.".

Additional reductions were proposed in 1997 by the House National Security Committee in H.R. 1778, which was described as follows:

> To accelerate the process of reform, the House National Security Committee reported H.R. 1778, the Defense Reform Act of 1997, to the House of Representatives. This bill pursues meaningful reform in three basic areas: streamlining the defense bureaucracy, improving defense business practices and adding a measure of common sense to the environmental regulations governing the Department's operations. Chief among the bureaucratic reforms are initiatives to reduce headquarters staffs by 25 percent and the defense acquisition workforce by more than 40 percent. According to the Congressional Budget Office, these reforms will save $15 billion over the next five years and an additional $5 billion each year thereafter without taking into account the additional potential savings resulting from the mandated increases in competition of defense support services.[8]

The defense acquisition workforce continued as a source of congressional oversight during the 105th Congress. The FY 1998 Defense Authorization Act (P.L. 105-85) required a 25 percent reduction in the number of personnel assigned to DOD management headquarters and headquarters support activities over 5 years; it specifically directed a 5 percent reduction during FY 1998, as well as a 5 percent reduction in staff at the United States Transportation Command during FY 1998. The compromise reached on the downsizing of the defense acquisition workforce (previously, the FY 1998 House Authorization Bill contained a provision that would have mandated a reduction of 124,000 personnel by October 1, 2001, but the Senate bill omitted any provisions) was to require a reduction of 25,000 defense acquisition workforce personnel in FY 1998; included in this bill are provisions that grant authority to the Secretary of Defense to waive up to 15,000 of the 25,000, based on his assessment that a greater reduction would "be inconsistent with cost-effective management of the defense acquisition workforce system to obtain best value equipment and would adversely affect military readiness."

The FY 1999 Defense Authorization Act directed the administration to reduce the workforce by 25,000 acquisition personnel by October 1, 1999, lowering it to 12,500 personnel if the Secretary of Defense certifies that such a reduction would cause an adverse effect on military readiness or management of the acquisition system.

THE FUTURE ENGINEERING FORCE

It is important that the Air Force evaluate its needs for scientists and engineers for the future, access them in proper numbers, develop and train them, and assign them to extract the best value for the Air Force. The development and training of these officers and civilians should include OJT that is supervised by

[8]House of Representatives Report 105-132, National Defense Authorization Act for Fiscal Year 1998, Report of the Committee on National Security, House of Representatives, on H.R. 1119 together with Additional and Dissenting Views, June 16, 1997, Washington, D.C.

qualified, experienced personnel; academic training when required to maintain and sharpen skills (much like the Weapon School does for operators); and education with industry.

The Air Force Academy programs for SE and systems engineering management (SEM) have important value; however, those programs will never produce graduates with these majors in large numbers. Although these students will not be qualified to be practicing systems engineers upon graduation from the Air Force Academy, their Air Force Academy training will have instilled in them an appreciation for what systems engineering means and for its importance. It is important that the Academy work with the Air Force Personnel Center regarding assignments of its graduates so that the Air Force can capitalize on the cadets' SE training. The Academy might benefit from an adjunct faculty position called Professor of Practice in Systems Engineering, similar to the faculty positions at USC mentioned earlier.

Similarly, AFIT's SE program will not graduate students trained in SE in large numbers. However, their training will be of value to the Air Force as it is applied in future assignments. AFIT might also benefit from an adjunct faculty position called Professor of Practice in Systems Engineering.

REVITALIZING THE ACQUISITION CORPS

Because of the dearth of Air Force acquisition programs that an Air Force officer or civilian will be involved with during his or her career, there are not many opportunities to gain insight and experience from OJT. As a means to provide such opportunities, the Air Force Secretary and Chief of Staff could establish a small mentoring group made up of retired Air Force general officers and civilians and representatives from industry and FFRDCs with credibility in acquisition; the establishment of this group would be a way to start the revitalization of the acquisition corps to a high level of excellence and to identify initiatives to accelerate effective, affordable combat capability to the field. This group would serve to (1) mentor the acquisition personnel, (2) provide individual and private expert advice and counsel to program managers, and (3) at the strategic level make recommendations to the Chief of Staff and Secretary on policy that could accelerate the revitalization process. This close and meaningful attention to the people of the acquisition community is called for in conjunction with the new policies coming from the Defense Acquisition Performance Assessment and the 40,000 person cut that the Air Force is currently mandated to take; its workforce needs to be better trained, more efficient, and more motivated than ever before. The military members of the acquisition workforce should deploy and serve as part of the warfighting Aerospace Expeditionary Forces if they are to be the greatest value to the Air Force during conflicts, and later when they take responsibility for acquiring future weapons systems and apply the valuable lessons learned in the field.

Technology and the producing of superior weapons systems have been the bedrock of the Air Force along with its people since 1947; the acquisition expertise that provided that capability in the past has eroded, and something needs to be done now to demonstrate to the people that serve the USAF in this area that they are important and vital to a successful Air Force; the acquisition community in turn needs to understand that "it's all about combat capability."

CONCLUDING THOUGHTS

Questions That Need to Be Addressed by DOD

All the military departments are wrestling with the role of SE in research and development (R&D) and the collective composition of the acquisition and science and technology (S&T) workforce for the Civil Service, the military, supporting FFRDCs, the service industry, and contractors. As SE matures as a discipline, terminology, curriculum, practices, and so on will become more standardized.

Developing an SE workforce will continue to be a challenge. The requirements in industry and government far exceed the number of qualified SE in the workforce. However, more fundamental philosophical questions should be addressed by DOD that will require hiring and/or retraining of engineers with new skill sets. These questions include the following:

- Should the DOD military service component be the lead systems integrator for large system-of-systems systems or should this role be contracted?
- Should the R&D structure in the laboratories be transitioned to one that is balanced in basic science, engineering, and SE competencies?
- Should SE be a recognized functional area within both the military and civilian workforce?
- What are the roles of the FFRDCs in oversight and research in SE?

Contractor and Government Considerations

The government and contractors each require experience in and access to similar or the same systems engineering competencies. Differences arise from the application and focus of these skills. The government's focus should be on developing requirements, on pre-Milestone A activities, and on monitoring and assessing the contractor's performance during pre-Milestone A and throughout programs through close coordination with the contractor(s). While the government is the "customer," the contractor(s) plays an important role in informing the government regarding what is possible or not. Government's challenges are to understand and manage programs and ensure that the contractors and the program offices have well-designed and fully integrated systems engineering plans (SEPs) and follow the documented processes.

Contractors often align their SE efforts based on functions that provide the engineering integration across the program life cycle. These functions need to be recognized and managed, particularly in the early phases of program planning. Some of the functions and skills (translated directly to job titles in many companies) that might be found where a strong emphasis on SE exists include these:

- Operations/systems analysis;
- System(s) architecture;
- Affordability analysis;
- Modeling and simulation;
- Integration, verification, and validation;
- Reliability, maintainability, and supportability;
- Human factors and ergonomics;
- Certification/qualification;
- System security;
- System safety;
- Integrated risk management;
- Testing and evaluation; and
- Configuration management.

FINDINGS AND RECOMMENDATIONS

Based on the discussion in this chapter, underpinned by the many briefings that the committee heard and by discussions within the committee itself, the committee offers the findings and recommendations shown below. They represent an accumulation of information and evidence, as opposed to conclusions of a particular specific study or studies.

Finding 3-1. The creation of a robust systems engineering process is critically dependent on having experienced systems engineers with adequate knowledge of the domain relevant to a contemplated program.

While the systems engineering process is, broadly, reusable, it depends on having domain experts who are aware of what has gone wrong (and right) in the past recognize the potential to repeat the successes under new circumstances and avoid repeating the errors.

Ideally, a person or persons with domain knowledge would have had experience working on exactly the same problem, or at least a problem related to the one at hand. If that is not so (and it might not be if the problem has never been addressed before, as was the case for Apollo and nuclear submarines), the term could be taken to refer to academic training in the relevant field of engineering or science. It would also refer to the practice in critical thinking and problem solving that comes with learning to be a systems engineer and then building on that foundation to gain the experiential knowledge and understanding of engineering in the context of an entire system. Systems engineering is enabled by tools that have been developed to assist in the management of systems engineering (not to be confused with the *practice* of systems engineering).

Both industry and Air Force presenters told the committee that there are not enough domain-knowledgeable and experienced systems engineers to support all of the programs that need them.

Recommendation 3-1. The Air Force should assess its needs for officers and civilians in the systems engineering field and evaluate whether either its internal training programs, which include assignments on Air Force programs that provide mentoring by experienced people and hands-on experience in the application of systems engineering principles, or external organizations are able to produce the required quality and quantity of systems engineers and systems engineering skills. Based on this assessment, the Air Force first should determine how and where students should be trained, in what numbers, and at what cost, and then implement a program that meets its needs.

The Air Force needs to attract, develop, reward, and retain systems engineers across the full spectrum of relevant domains, engage them in the early (pre-Milestone A) phase of new programs (or modification programs), and sustain their participation throughout the life of the programs. One important step in this process would be to create an Air Force occupational code for systems engineering so that engineers' experience and education can be tracked and managed more effectively. The Air Force should support an internal systems engineering career track that rewards the mentoring of junior systems engineering personnel, provides engineers with broad systems engineering experience, provides appropriate financial compensation to senior systems engineers, and enables an engineering career path into program management and operations.

Finding 3-2. The government, FFRDCs, and industry all have important roles to play throughout the acquisition life cycle of modern weapons systems.

Since the need for a new or upgraded weapon system is most often first recognized by the military user, it is appropriate for the military to codify its requirements and, with support from FFRDC and independent systems engineering and technical assistance (SETA) contractors, to explore materiel and nonmateriel solutions (such as doctrinal, organizational, or procedural changes) as well as to assess the potential for new technology to provide enhanced capabilities. While it is appropriate and usually desirable to engage development contractors in the pre-Milestone B process using competitive study contracts, the source selection for system development and demonstration should not be made until after the work associated with Milestones A and B is complete.

Recommendation 3-2. Decisions made prior to Milestone A should be supported by a rigorous systems analysis and systems engineering process involving teams of users, acquirers, and industry representatives.

Working together, government and industry can develop and explore solutions using systems engineering methodology to arrive at an optimal systems solution.

4

Systems Engineering Functions and Guidelines

INTRODUCTION

In this chapter the committee defines the most critical systems engineering (SE) activities in the pre-Milestone A/B phase. Six key functions are discussed, along with four key SE outputs for the Milestone A/B activities. The committee does not attempt to capture every function that may be involved, but instead focuses on the broad functions that have the greatest impact on ultimate system performance. The intent is that the chapter highlight the key functions from the viewpoint of the program manager, not define a "how to" manual for systems engineers.

The committee discusses guidelines for managers and systems engineering practitioners in the pre-Milestone A/B phase of a program. In preparing these guidelines, the committee identifies those items that address the causes of the large decline in development program productivity and the attendant cost and schedule growth over the past 30 years, as discussed at the beginning of Chapter 1. These guidelines cover the following topics:

- *The "six seeds of failure":* a discussion of six areas of risk that the committee believes are worthy of particular attention in the pre-Milestone B phases and that are the basis for the recommendations that the committee makes later;
- *Other potential causes of development productivity decline:* a discussion of several often-mentioned causes of development productivity decline, including talent diminishment in government and industry, and excessive oversight;

- *Obsolete and nonrelevant systems engineering processes:* a discussion warning against the use of legacy processes and guidelines that do not add value to the tasks at hand;
- *General processes and practices for systems engineering:* a discussion of some of the processes that are critical after Milestone B, including modeling and simulation, control of lower-level requirements, and change control; and
- *A checklist:* some of the committee's most important findings, provided to guide systems engineering activities and output before Milestone B.

PRE-MILESTONE A SYSTEMS ENGINEERING FUNCTIONS

The prerequisite for starting the systems engineering process is a user-defined need (or outcome). The purpose of systems engineering in this phase is to define an optimal system concept and concept of operations (CONOPS) that satisfies the need and also fits within budgetary constraints. The systems engineering outputs in this phase include a concise statement of key performance parameters (KPPs), a CONOPS description, a program implementation strategy, and models to support performance assessment, cost estimation, and risk management in subsequent phases.

These functions are aggregated in the following broad areas, shown in Figure 4-1: concept creation, preliminary CONOPS, performance assessment, architecture development, risk assessment, and cost scoping. These functions are reiterative and thus occur in parallel, to produce the key outputs from the pre-Milestone A systems engineering process: KPPs, CONOPS, supporting models, and program implementation strategy.

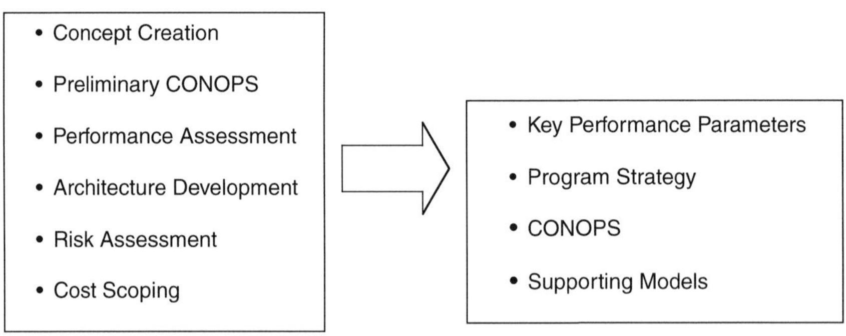

FIGURE 4-1 Critical pre-Milestone A systems engineering functions and outputs. NOTE: CONOPS, concept of operations.

Functions

The functional elements are discussed briefly below.

1. *System concept creation.* This is a creative process for selecting a basic approach for a system to achieve the desired outcomes. The following guidelines are important in this phase:
   - *Consider alternatives.* It is important that at least two alternative concepts be considered before locking in a solution.
   - *Consider the "time value of capability."* Seek concepts that can deliver initial capability within about 5 years (measured from Milestone B). Each year that a needed capability is delayed has a cost to those who need it, and delays the availability of operational data and experience to guide subsequent improvements. Development programs that are successful and timely at initial operational capability (IOC) often become platforms on which many other capabilities are subsequently added. The world changes too fast to be friendly to long development cycles. Further, extended delivery times run the risk of the system becoming obsolete before deployment and can be an indication that the concept is excessively complex or excessively dependent on immature technology in its first delivery. Indeed, one insidious effect of long development cycles prior to IOC is that they create the temptation to add new emerging technologies and to further tune the requirements, which have the effect of further increasing the pre-IOC development cycle. Historical precedent shows that many complex major systems can achieve IOC in less than about 5 years.

The Air Force has a detailed analysis of alternatives (AoA) process that can be invoked in this phase. In Air Force Instruction (AFI) 10-601,[1] the role of the AoA process is described as follows:

> The AoA provides information that helps the decision makers select the most cost effective alternative(s), in order to satisfy a mission need or eliminate an operational gap/shortfall in capability. It compares alternative solutions on the basis of operational effectiveness, and cost. It documents the analytical and operational rationale for choosing the preferred alternative(s). It also helps to justify the need for starting, stopping or continuing an acquisition program.

In its grandest form, this process is extensive in defining methodology and documentation requirements and should be tailored to meet the needs of each program.

[1]Office of Aerospace Studies Analysis of Alternatives (AoA), 2003, *Guidance in Support of AFI 10-60* (revised September 22, 2003), Kirtland Air Force Base, N.Mex.: Office of Aerospace Studies.

2. *Develop a CONOPS.* At this stage, the CONOPS is a top-level description of how a system and its operators and users will interact to produce the required capability, including operation in a system of systems. At Milestone A, the CONOPS needs to be sufficient to ensure that the chosen concept is capable of operating to produce the desired outcomes and time lines. A key precursor of a CONOPS is often a description of the other systems with which the system of interest will be interacting and an understanding of what the user expects the system to do in its operating environment.
3. *System performance assessment.* This is the analytical process of predicting the performance of the system concepts evaluated. It needs to be sufficient to predict the ultimate performance of the system in operation. Often system performance models developed in this phase become the basis for supporting the government's management of the development phase. The following guidelines are important to consider:
    - The models must typically also provide the basis for setting segment performance requirements and assessing the impact of deviations from expected performance after Milestone A.
    - Modeling and simulation often constitute a powerful tool, starting with the evaluation of concepts and moving on through the entire development phase and even into the operational phase. While this discipline may not always begin in the pre-Milestone A phase, models begun in this phase can add rigor to the assessment of concepts as well as being powerful management tools in subsequent phases.
4. *System architecture development.* A well-known approach to addressing a complex problem is to break it into parts that can be addressed separately. Architecture here refers to the partitioning of the system into separately definable and procurable parts, the structuring of interfaces between the system and the outside world, and the structuring of interfaces (physical, functional, and data) among the segments. Through careful partitioning, architecture can minimize complexity and thereby development risk. The following guidelines are important to remember in this phase:
    - *Seek independence of the segments.* Good architecture seeks to achieve segments that can be developed and tested separately to reduce the complexity of the development phase.
    - *Simplify interfaces.* In selecting the architecture, seek to minimize the entropy of the implementation by making the interfaces among the segments as simple as possible.
    - *Plan for testability.* It is important that the system designer consider the testability of the parts alone, of the system as a whole, and of any system of systems implications as early as possible to minimize the impact of undiscovered issues later.
    - *Plan for integration.* The system designer must have a vision of how the parts can be integrated and tested to verify end-to-end performance

in the operating environment. Modeling and simulation can often play a major role in the overall test and verification plan.

5. *Risk assessment.* A pre-Milestone A goal is to identify the key risk drivers to set the stage for effective risk management in the procurement and development phases. The six key risk drivers identified in the next section are areas to focus on in risk assessment. High-risk areas may be found not only in technical areas, but also in the availability of critical human resources, stable funding, industrial base capacity, and so on.
6. *Cost scoping.* The systems engineering process in the pre-Milestone B phases normally includes very rough estimates of a system's development and operating costs sufficient to determine if the concept and CONOPS are consistent with expected budgetary constraints. The system designers should try to avoid the low bias in cost estimating often introduced by the desire of the government and potential contractors to sell the program. This is usually best accomplished, in this early phase, by top-level comparisons on a segment-by-segment basis with actual cost experience on recent programs. For revolutionary concepts, less-well-grounded techniques may be unavoidable. Models developed in this phase may be matured later to support "should costs" in evaluating contractor bids.

Outputs

The output categories are described briefly as follows.

1. *Definition of key performance parameters.* This is the primary output of systems engineering in the pre-Milestone A period. These KPPs will drive the subsequent procurement and development stages of the program. It is important that they are defined concisely and specifically. They should be few in number, but specific in describing the primary capabilities desired. They should be sufficient to provide the basis from which all lower-level requirements are derived. Ideally, they are defined in terms that are understandable to the users of the system rather than in highly complex technical language. In the committee's experience, the failure of the designers to define the system's KPPs simply and clearly at Milestone A and the top-level requirements at Milestone B is the first step to requirements instability and overruns later.

   The committee believes that the top-level KPPs and system requirements can be the basis for all lower-level requirements. Committee members have had experiences with complex, multisegmented systems where a half-dozen top-level requirements drove all lower-level requirements. Often, system availability, maintainability, and/or reliability requirements provide the basis for important derived requirements that are not covered by other capability requirements. The imposition of this discipline can

TABLE 4-1 Summary of Pre-Milestone A Systems Engineering Function and Output Guidelines

Function/Output	Summary Guideline
System concept development	• Complexity should be minimized both within the system and with regard to the system's external interfaces. • The use of high-risk, immature technologies should be avoided. • Favor concepts that can achieve initial operational capability (IOC) in fewer than about 5 years. • At least two alternative concepts should be evaluated before selecting a final concept.
System architecture	• Partition to achieve segments that can be procured and tested separately. • Minimize the complexity of the interfaces among the segments. • Establish data, structure, and architecture standards where appropriate. • Maintain the independence of the functional requirements.
Concept of operations (CONOPS)	CONOPS needs to be developed sufficiently to ensure that the concept can: • Perform in the environment in which it will operate, • Handle the expected throughput, and • Meet response time requirements.
Performance analysis	Performance analysis should be sufficient to: • Predict performance against mission needs, • Assess the impact of segment-level performance on end-to-end performance, and • Include the development of system performance models to support performance assessment in the acquisition phase.
Top-level requirements generation	Requirements at this stage should: • Be broad and few in number, • Drive the derivation of all lower-level requirements on the program, and • Remain largely unchanged through the first development cycle and the achievement of IOC.
Risk identification	Risk identification should: • Identify the top-level risk factors that are inherent in the concept, architecture, and CONOPS.
Cost estimates	Cost estimates should: • Provide a high level of confidence that the concept and CONOPS are consistent with budgetary constraints, and • If necessary, facilitate the development of a cost model that can be extended later to support "should costs" for contractor-proposed solutions.
Key performance parameters (KPPs)	KPPs at this stage should be: • Broad and few in number; • Comprehensive and clearly and simply defined covering required capabilities, availability, and reliability; • Sufficiently complete to be the source of all lower-level requirements; and • Sufficiently mature that little change is needed after Milestone B, and prior to IOC.

have a powerful effect in protecting against requirements creep driven by well-meaning specialists who want to insert their favorite processes and technologies into the program.

The Department of Defense (DOD) *Guide for Integrating Systems Engineering into DOD Acquisition Contracts, Version 1.0*[2] has this to say on the importance of getting the requirements right:

> Sound system requirements (including performance) are the backbone of a good technical strategy and resultant plan (as documented in the SEP and related plans). The performance requirements, as a minimum, must be commensurate with satisfying the threshold for the critical operational (including sustainment and support) requirements (e.g., Key Performance Parameters (KPPs)) and balanced with program cost, schedule, and risk constraints. If these elements are not balanced at the start of the SDD phase, the program has a high probability of incurring cost increases, suffering schedule delays, and/or deficient performance of the end product.

2. *CONOPS*. A draft CONOPS is needed to show that the system can be operated to meet the users' objectives.
3. *Supporting models*. Models to validate system performance and cost projections in phases A and B are important to support the government's management of subsequent stages.
4. *Program implementation strategy*. The final output from a typical Milestone A activity should be a strategy for implementing the program going forward. This typically would include a plan and time line for achieving Milestone B, an approach to mitigating the most critical identified risks, and an approach to establishing cost projection credibility before Milestone B.

Table 4-1 summarizes some of the committee's observations on the key functions and outputs.

SIX DRIVERS OF COST, DEVELOPMENT TIME, AND PERFORMANCE RISK THAT ARE ADDRESSABLE BY SYSTEMS ENGINEERING PROCESSES

The committee heard wide-ranging views from its guest speakers, members, and others about the twofold-or-greater growth in program cost and completion time over the past 30 years. The committee believes that understanding the real drivers of this large increase in development times and costs can help identify

[2]Department of Defense, 2006, *Guide for Integrating Systems Engineering into DOD Acquisition Contracts, Version 1.0*, December 11. Available at http://www.acq.osd.mil/se/publications.htm. Last accessed on June 20, 2007.

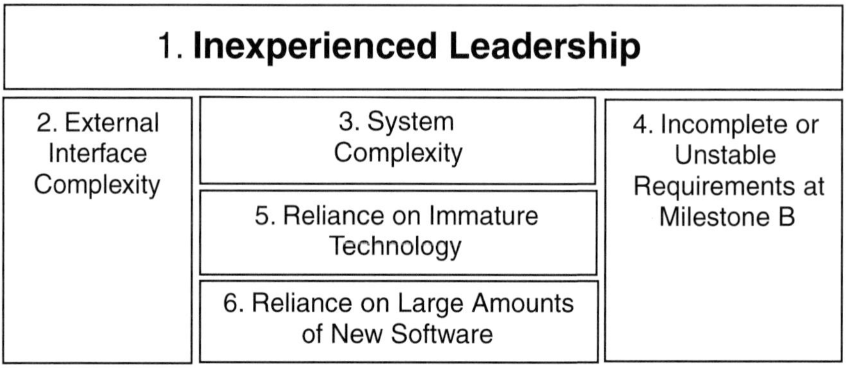

FIGURE 4-2 The "six seeds of failure" to avoid or manage during the pre-Milestone A and B phase.

the role that early applications of systems engineering processes could play in reversing this trend.

The committee's approach was to harness its collective experience along with the lessons derived from the case studies in Chapter 2. It also tested hypothetical drivers of the problem by comparing current practices with those that existed in the past when some of the most remarkable program successes were achieved in fewer than 5 years from Milestone B. While this was at best an anecdotal analysis, the committee examined the myriad factors that may have contributed to serious development issues and identified six factors that are pervasive sources of poor performance and that are addressable through sound systems engineering processes. The committee calls these factors, illustrated in Figure 4-2, the six seeds of failure. They are inexperienced leadership; external interface complexity; system complexity; incomplete or unstable requirements at Milestone B; reliance on immature technology; and reliance on large amounts of new software.

A brief description of the rationale behind each of these factors follows.

Inexperienced Leadership

Perhaps the biggest risk of all in undertaking large development programs is to proceed with less than the best personnel, particularly in the key leadership positions in government and industry. High-quality program managers and system engineering leaders, in particular, are critical. High aptitude and extensive experience, combining to create high domain knowledge, are required for individuals to be fully effective in these positions. In evaluating program management and system engineering experience, one should consider both the length of experience and the number of programs that the candidates have participated in. Experience in at least two major development environments can be very valuable. Since not every program can have the best, it is important for government executives to assign the best to the most difficult problems. Programs that exhibit the highest levels of the risk factors described here, particularly high complexity, are good candidates for the best leadership team.

Pre-Milestone A Mitigation

For programs with high levels of external interface complexity and internal technical complexity, the government should seek the highest-aptitude, most-experienced people it can muster for the program manager and system engineering manager roles going forward into the Pre-Milestone B phase, and it should keep these people in place through IOC whenever possible. Proven strength and the ability to minimize complexity and control changes after Milestone B through IOC in the face of unrelenting pressure both from inside the program team and from external users and sponsors, are important.

External Interface Complexity

One characteristic of very complex system developments during World War II and the early Cold War years was the simplicity and urgency of the needs and missions. Beating the Germans to the atomic bomb, the Russians to the Moon, penetrating the Iron Curtain, and so on, provided clear, urgent goals that galvanized the sponsors of the complex systems and focused and empowered the government and contractor teams. Such clear, driving missions, and the simple user interfaces that they required, allowed the program team to develop its concepts quickly and to keep the top-level requirements stable until IOC.

In the post-Cold War era, the immediacy of the threats often seems less apparent, and programs often try to serve many missions and users with a single system, or system of systems. The interaction of multiple systems that were not designed together (e.g., military satellite communications [MILSATCOM], see Chapter 2), often termed "systems of systems," also can greatly increase the difficulty of creating a stable requirements base for a new system. The concept of network-centric operations, in particular, can introduce external complexity. The complex processes necessary to coordinate these communities of interest seem to have hidden costs that can add many years to the development cycle and lead to substantial budget overruns. In addition, systems dependent on highly complex external interfaces can be far more difficult to operate after deployment.

Pre-Milestone A Mitigation

Systems engineering should treat the minimization of external interface complexity as a key driver in selecting concepts and architecture. Simplifying and standardizing the ways that external users access the system and seeking to minimize the degree to which the system's capabilities must be tailored specifically for individual users can help. Program managers may also be able to minimize the pressure for pre-IOC changes from the external user community by striving to create a sense of time urgency to aid in stabilizing requirements.

System Complexity

The flexibility and capability enabled by advances in electronics technology and software provide systems designers with far more options than their predecessors enjoyed 30 years ago. The downside is that these new capabilities can tempt designers into unnecessarily complex concepts and designs that impose a "cost of internal complexity" similar to the external complexity costs described above. In particular, there can be a tendency to assign poorly-thought-out functions and options to the software. Often, late-arriving requirements are implemented in software because the perceived schedule risk of redesigning hardware is unacceptable.

Pre-Milestone A Mitigation

Systems engineering should treat complexity minimization as a key driver in selecting concepts and architecture. Architecture selection can have a powerful effect on controlling complexity. A proven approach to solving large, complex problems is to partition the problem into smaller pieces that can be addressed separately. Partitioning the system to create parts that can be separately developed and tested and that minimize the complexity of internal and external interfaces can be very important.

Incomplete or Unstable Requirements at Milestone B

It is not unusual for programs to proceed beyond Milestone A with unsettled KPPs and beyond Milestone B with unsettled top-level requirements. This can be a major schedule and cost driver, particularly after Milestone B. As the missions and user communities have become more complex, the government is finding it much more difficult to resolve competing views of system concepts and performance requirements. As a result, it is often tempting to go through the process of selecting the development contractor team before resolving these issues. The well-meaning goal is to get the contractor team's help in resolving these issues. The result, however, is often that large development teams start without sufficient direction, and the contractor teams with their high "burn rates" waste time and money participating in "what-if" requirements debates that do not converge quickly.

Another driver of unstable requirements is funding instability. This phenomenon, familiar to all, is a result of political processes, not systems engineering.

But its effects on requirements stability, cost, and schedule are significant and insidious. A committee member told of his experience with the Defense Satellite Communications System (DSCS) Program and how the rate of design and production accelerated dramatically, and how the costs dropped significantly below expectations when a fixed-price, multiyear procurement contract was authorized by the Congress and, in effect, froze funding and changes to the requirements.[3]

Pre-Milestone A Mitigation

Federally funded research and development centers (FFRDCs), systems engineering and technical assistance (SETA) contractors, and small study contracts to potential development contractors can all be used to resolve concept and requirements issues prior to Milestone A. It is important not to proceed with the development phase until the top-level concept and requirements are finalized. Also, tight partnership and collaboration between the users/sponsors and the developer are critical to stabilize requirements in the formative phase of a program.[4]

Reliance on Immature Technology

The use of unproven technology in large system developments can introduce a high risk of schedule and cost growth. The committee has seen examples of large procurements in which high-risk technology was seen as a part of the rationale for justifying a program. The committee believes that the use of mature technology should be a prime goal in concept and architecture selection. When the concept and architecture selections identify technologies that are risky, alternative concepts should be found, or the program strategy must accept that the system must wait for the technology to be demonstrated successfully before entering system design and development.

Pre-Milestone A Mitigation

The system designer should seek to avoid the use of new, unproven technology in large development programs with important missions. The committee believes that technology development should be conducted separately from large operational program developments if possible.

As an alternative to pursuing such technology in parallel with post-Milestone B development, the systems engineers might consider making provisions in the architecture and systems design at Milestone B to allow subsequent technology

---

[3]Lt Gen Ronald Kadish (USAF, ret.), Booz Allen Hamilton, "Defense Acquisition Performance Assessment: Report Summary Briefing," presentation to the committee, February 28, 2007.

[4]National Research Council, 1993, *An Examination of the Air Force's Pre-Milestone One Planning/Decision Process*, Washington, D.C.: National Academy Press.

insertions of expected new technology, if such flexibility can be accomplished without adding substantial design complexity or requirements instability.

Reliance on Large Amounts of New Software

Large, complex software elements have been the source of high costs and long development times on many programs. The committee believes that most often this is not due to the difficulty of the software design, but rather reflects an inadequate definition of the software requirements prior to initiating the software development effort. Poorly-thought-out and ambiguous requirements seem to be an even greater addiction in software than in systems as a whole. A common view is that "after all, we can always change it later." The tendency to accept far too much complexity in software requirements is also a common problem.

Pre-Milestone A Mitigation

Perhaps the most important actions that the systems engineers can take to minimize the risk of software disasters are to get the software functional allocations clearly defined at Milestone B and to constrain the software such that the software element can be delivered within about 18 to 24 months or less. In 2000, the Defense Science Board[5] stressed the following guidelines for software program structure that the committee believes are worthy of attention:

- Aggressively limit development time to no more than 18 months;
- Minimize complexity;
- Highly incentivize development;
- Allow program management to trade functionality for time and stability;
- Have good processes, but value past performance over process;
- Use an iterative, not a waterfall, development process; and
- Develop an executable architecture first.

Contrary to intuition, the integration of commercial off-the-shelf (COTS) products often results in high technical risk in DOD acquisitions. COTS integration needs to be planned and assessed in the overall system context. Performance issues in the context of the overall system architecture need to be taken into account. Any modification to COTS software usually implies life cycle maintenance of the entire COTS product. Integration issues are often more challenging than in commercial applications. With reference to the "Pre-Milestone A/B Checklist" (Box 4-1, which appears below in the section entitled "Pre-Milestone A/B Checklist"), early consideration should be given to COTS risks. Sources of

[5]Defense Science Board (DSB), 2000, *Task Force on Defense Software*, November, Washington, D.C.: OUSD (AT&L). Available at http://www.acq.osd.mil/dsb/reports/defensesoftware.pdf. Last accessed on June 27, 2007.

useful information tailored to the acquisition community, including known risks, publications, case studies, courses, and a risk assessment tool called the COTS Utilization Risk Evaluation, are maintained by the Software Engineering Institute (see www.sei.cmu.edu/cbs).

OTHER POSSIBLE PERFORMANCE DRIVERS

In addition to the six seeds of failure discussed above, committee members and others discussed a number of other factors as potential sources of the decline in development productivity over the years. Four of these were most frequently mentioned: diminished talent in the government, diminished talent in industry, excessive focus on cost and schedule to the detriment of performance, and excessive oversight. While it may be correct to identify each of these factors (in particular the diminished talent and domain experience of key personnel), the committee believes that the others may be less significant than many believe and that they are certainly not excuses for poor performance or reasons not to look for solutions in good systems engineering and program management. Table 4-2 briefly summarizes committee observations on these often-cited issues.

TABLE 4-2 Other Cited Sources of Development Productivity Decline

Source of Decline	Committee Observation
Diminished talent in government to manage acquisitions	It is not clear that the talent in the government is less capable than it was 30 years ago. The "good old days" may not have been as good as we think they were. In fact, in some ways the depth of talent today is greater owing to the entrance of women and minorities into the available talent pool, and the much larger and more experienced government-industry talent pool to draw from. But there are issues involving the number and the domain experience of available personnel in government.
Diminished talent in industry	Same observation as above.
Focus on cost and schedule to the detriment of performance	The committee believes that programs benefit from strong commitment to cost and schedule, as well as to performance and supportability.
Excessive oversight	This may be the most important of these issues. The committee believes that the most insidious effect of excessive oversight may be its contribution to requirements and funding instability between Milestone B and initial operational capability. Another negative effect is the tendency of high-level external "oversight" to defocus the management team from the development job and direct its energy instead to providing air cover for the program.

OBSOLETE AND NONRELEVANT SYSTEMS ENGINEERING PROCESSES

At least one major prime contractor known to the committee has decided to eliminate the term "systems engineering" altogether after finding that many of the accumulated documented processes in government, academia, and industry are useless. Processes can be a valuable tool to capture the lessons of the past. However, in the committee's experience, the accumulation of such processes over many years, largely implemented to address one specific development problem or another, can lead to programs being driven by rules that are irrelevant, obsolete, or excessive for the real-world job. Process requirements generated by well-meaning people who do their work without the benefit of real program experience can also lead to non-value-added work. The adverse effects of obsolete and nonrelevant process requirements can be minimized by allowing systems engineering and program management the leeway to tailor compliance with required processes to suit the needs of each specific program.

GENERAL POLICIES AND BEST PRACTICES FOR SYSTEMS ENGINEERING IN ALL PHASES

The previous sections of this chapter concern guidelines and practices for program-specific systems engineering efforts during pre-Milestone A and B phases. This section addresses some of the most important general guidelines and practices for personnel and organizations responsible for systems engineering policy, methods, and tools that apply to the entire program life cycle.

The prime contractor assumes much of the responsibility for systems engineering after Milestone B, with the government involved in oversight, review, and approval. Some of the pre-Milestone A functions are complete (such as concept development). Others, such as architecture and CONOPS development, move to a more detailed level. New functions come into play, such as change control, configuration control, interface definition and management, and detailed modeling and simulation of operational scenarios. Avoiding requirements creep as low-level requirements are developed by the contractors is an important government systems engineering role after Milestone A. The following paragraphs discuss the committee's views on several key systems engineering functions after Milestone B.

Modeling and Simulation

Modeling and simulation are major components of the systems engineering process throughout the life cycle of a system. As a system moves through the acquisition cycle, modeling and simulation transition from more aggregated and general forms to higher-resolution and higher-fidelity forms. Early in the cycle, constructive models tend to be prevalent, while later the activity shifts toward human-in-the-loop and mixed-mode simulation to support operational testing and evaluation. As a system moves into operational use, live rehearsal may become the most prevalent form in use.

It is now possible to represent varying portions of a system with simulations throughout the development, integration, and testing processes even before real hardware and software become available, making it unnecessary to wait for the completion of development before any substantial integration testing can be done.

During the early stages of planning for a new program, management should recognize the need to acquire or develop appropriate data necessary to enable the timely use of modeling and simulation prior to Milestone A.

Another challenge that must be addressed is that of maintaining consistency across the different levels of modeling and simulation aggregation and fidelity. Moving from the low- to medium- to high-fidelity models, the outputs of one stage will generally serve to set performance thresholds for the next stage, helping to define what constitutes a successful concept or design.

The potential bidders for the systems to be acquired will also play a significant role in defining the simulation environment and by building contractor-specific models expressing both the behavior and performance of their concepts.

At each level of aggregation and for each iteration of the models, there will be systems engineering tasks to validate both the models and the environment, ensuring that the laws of physics are taken into account and that behaviors are both consistent and explicable.

Modeling, simulation, and analysis will play a major role in identifying key design parameters—the "long poles in the tent"—which are associated with potential risk, sensitivity, or uncertainty and on which the ability of the design to meet important requirements will depend. Whether the parameter is processing throughput or response time or the acquisition cost per removal-free operating hour, early identification provides guidance to the designers and focus to risk-mitigation planning activities such as early benchmarking or development of alternatives.

Many program managers fail to recognize the return-on-investment of planning for and using modeling and simulation up front in the systems engineering process, and therefore they reduce or eliminate the budget for modeling and simulation from their overall budgets. A relatively minor up-front investment in modeling and simulation can actually reduce significantly the cost of the development program by identifying problems early and reducing both the time and the cost of full-scale testing.

Modeling and simulation have become ever more central to the development of modern systems. Unprecedented advances in digital processing have made high-fidelity representation of systems and subsystems in computer models possible from the simplest of systems to the most complex. This has made it possible to examine the projected performance of systems over wide excursions of design and environmental assumptions very early in the development process, even prior to Milestone A. Today's modeling and simulation tools make it possible to perform extensive system of systems simulations and evaluate alternate architectures at affordable cost and early enough to make a difference.

Systems of Systems

The committee had numerous discussions on "systems of systems" (SoS) and found the term ill-defined and overused. However, there are some important issues that arise when a system designed to provide a primary capability is used in conjunction with other systems to provide a different capability.

The definition of SoS that the committee adopted is as follows: systems of systems are groups of systems, each of which individually provides its own mission capability, that can be operated collectively to achieve an independent, and usually larger, common mission capability.

Often systems of systems are not initially developed together but rather are formed by operating systems, initially developed separately into a SoS to achieve a larger objective. The Army's Future Combat Systems (FCS) Program is an example of a SoS in which the constituent systems are being procured together. This SoS consists of platforms, weapons, surveillance systems, and so on that can operate separately or together in combat.

While the design of systems of systems may impose somewhat different engineering processes, the committee believes that, in general, most of the elements of good systems engineering apply to systems of systems as well, and the committee has not attempted to differentiate between them in this report. However, these types of system aggregations increase the complexity of external interfaces that are difficult for a development program office to define and control. As pointed out earlier, high levels of external interface complexity constitute one of the six seeds of failure and deserve specific attention. Thus the use of memorandums of understanding and other cross-organization means of communication and control are particularly important in these types of programs.

System Dynamic Modeling

System dynamics seeks to understand, through qualitative and quantitative models, the time-dependent behavior of managed systems and how information feedback governs their behavior. System dynamics also seeks to design robust information feedback structures and control policies through simulation and optimization.

System dynamic modeling[6] has the potential to allow development program performance to be predicted with far greater accuracy by modeling interactions

[6]K.G. Cooper, 1994, "The $2,000 Hour: How Managers Influence Project Performance Through the Rework Cycle," *Project Management Journal* 15(1):11-24.

among internal and external social and environmental factors. There is evidence that it is possible to model a program based on its organization, management policies, staffing, experience level by skill, productivity, overtime, attrition, morale, hiring, training, requirements changes, quality, required rework, and so on, and to calibrate the organization's processes with historical data. The resulting system dynamics model could be used to help establish appropriate policies and guide program management to insightful decisions that affect program success. While the committee did not hear any accounts of the successful use of system dynamics modeling on government development programs, the committee believes that this discipline deserves further evaluation as a systems engineering and program management support tool.[7]

While system dynamics modeling is relevant to both the system being developed and the development process itself, there are many equally good (some would argue substantially better) modeling methods that can deal with the product being acquired. Discrete time simulation lends itself to modeling systems governed by physics, and discrete event simulations tend to be used for systems governed by statistical processes. System dynamics seems uniquely relevant to pre-Milestone A activities in modeling the process of acquisition. The challenge during the pre-Milestone A period is to capture the potential interactions between the organizational, political, economic, and policy components of the acquisition system as much as it is to capture the interactions between the components of the system to be acquired. System dynamics models might provide great value, and their power to stimulate insight might lead to significant advances in the practice of systems acquisition.

Testing and Evaluation

The discussion of modeling and simulation above can be expanded to include the total testing and evaluation (T&E) function. Testing and evaluation could and should be used to validate key outputs of pre-Milestone A systems engineering. This requires coordination between the planning and systems engineering community and the T&E community, to ensure harmony and linkage between what is tested and validated and what is defined in the front end of programs. Indeed, one of the best forcing functions to get people to agree on what is expected is to get detailed agreement on the tests that, if passed, demonstrate a successful completion. Taking the position that no requirement exists until the test that it must meet has been defined is a valuable tool in managing successful projects, in the committee's experience. Also, the smart linkage between the planning and testing communities, including the wise application of modeling and simulation, can actually reduce the need for some costly, time-consuming, full-scale testing.

---

[7]Professor Jay W. Forrester of the Massachusetts Institute of Technology has been a prominent researcher in this area. Further discussion of the topic can be found at http://sysdyn.clexchange.org/sd-intro/home.html. Last accessed on November 20, 2007.

The committee believes that a review of how T&E is incorporated in the development planning process, how modeling and simulation can be used effectively in T&E, and how T&E supports program validation can be of immense value.

Cost and Schedule Performance Estimating

Large cost and schedule overruns on development programs have become the norm, and they continue to seriously undermine confidence in the acquisition process, stress the nation's resources, and diminish the capability of the nation to respond quickly to new threats. Cost and schedule results are heavily impacted by SE decisions in the pre-Milestone B phases, and subsequent performance cost and schedule efficiency can be significantly affected by the application of good SE discipline during the development phase.

The two, cost and schedule performance, are almost always inextricably linked. That is, the causes of cost deviation affect schedule, and vice versa. These causes can be discussed in three groups: (1) those that are the result of program *execution performance*, (2) those that result from *technical estimating uncertainty*, and (3) those that result from *competition-driven biases* in the initial cost and schedule estimates.

By "execution performance," the committee refers to how closely ultimate cost and schedule performance mirror the cost and schedule performance obtainable by an optimally managed program. Since optimum management on a development program is not possible, the variances here are always negative (overruns) that must be offset by recognizing this fact in the initial estimates.

This report has attempted to identify guidelines for systems engineering on development programs that can enable the programs to be executed as closely as possible to this theoretical optimum. These guidelines are summarized in this chapter and in the committee's checklist (see the section below titled "Pre-Milestone A/B Checklist"). Elsewhere in the report, the committee also addresses Air Force-wide issues—such as the training of system engineers and certain Air Force organizational issues covered in the committee's recommendations—that can better enable these guidelines to be executed on all programs. In comparing today's program performance to that of similar programs of years ago, the committee examined the actual Milestone B-to-IOC times rather than performance-to-budget, so that it would be comparing the effects of program execution, not the estimating errors discussed below.

By "technical estimating uncertainty," the committee means the uncertainty about exactly what must be done, how long a development will take, and how much it will cost, given the fact that the job has not been done before. One might assume that this uncertainty would have a zero mean, that is, that error would sometimes be too high and sometimes too low. But in real life, the committee's experience is that this uncertainty almost always leads to a low estimate of the cost and schedule required. This may be because it is unconsciously assumed that the program execution will be optimum and that all of the tasks to be done have been identified. In real life, execution can never be optimum, and at the outset there is always a failure to identify a significant fraction of the work needed. Good program managers know this and build contingencies into their cost and schedule estimates.

By "competition-driven biases," the committee means the desire of government and industry supporters of a program in the pre-Milestone A and B phases to sell the program to the executive and legislative branch budget officials and of the competing contractors' desire to win. The competing contractors, for what are usually cost-plus contracts, are highly incentivized to bid the "lowest credible cost" to win—as one said, "It's better to be on contract and underbid than be on the street." A now-retired senior aerospace industry executive is reported to have told one of his losing teams, "There is no excuse for losing a cost-plus procurement on cost." The result of government's and industry's desire to sell their programs and contractors' desire to win creates a "conspiracy of hope"—an expression given some prominence by the Defense Acquisition Performance Assessment (DAPA) panel.[8]

It should be noted that the estimating errors created by the second and third issues above cause a program to be planned with cost and schedule objectives that are not achievable, and that cause significant energy to be expended by government and contractor management teams as the variances are recognized and plans painfully redone.

The systems engineering process can play a role in mitigating these factors by developing cost projection models during the pre-Milestone A and B phases. These models seem to work best when they are rooted in real-life cost experience on similar developments, with adjustment made for differences in complexity and novelty in the programs being compared. Success-oriented schedules, with insufficient time and budget margins for the unknown development problems that are sure to come, are a classic attribute of underestimated programs. The committee believes that systems engineering and program management on every large development program should develop a cost model and discipline late in the pre-Milestone A phase and refine the model through Milestone B.

Requirements Creep and Requirements Traceability Matrices

After Milestone A and to a greater degree after Milestone B, the contractor, with help from government overseers, will be developing detailed requirements at the segment, subsystem, and component levels. Systems engineering in govern-

---

[8]Ronald Kadish, Gerald Abbott, Frank Cappuccio, Richard Hawley, Paul Kern, and Donald Kozlowski, 2006, *Defense Acquisition Performance Assessment*. Available at http://www.acq.osd.mil/dapaproject/documents/DAPA-Report-web/DAPA-Report-web-feb21.pdf. Last accessed on April 2, 2007.

ment and industry must control this process so that each lower-level requirement is directly responsive to top-level system requirements or KPPs.

There is a strong tendency for subsystem and component engineers to expand the cost and development difficulty by writing specifications based on the "best the supplier can do," or on personal preference, rather than on what the system needs. This ratcheting up of requirements can have unexpected cost and schedule impacts. Requirements traceability matrices can be useful tools for making sure that each lower-level requirement can be justified by a system-level requirement. This helps enforce the discipline that all lower-level requirements must be tracable to one of a few well-thought-out KPPs and top-level system requirements discussed earlier in this chapter.

Change Control and Configuration Management

Change control becomes a very important activity after Milestone B. It plays a critical role in guarding against the top-down and bottom-up requirements creep and requirement instability, discussed above with respect to the six seeds of failure. During the pre-Milestone A period, the KPPs of the system (or system of systems) should be established and agreed to by the user/sponsor, the developer, and senior management authority. These KPPs should be placed under tight change control at Milestone A. Thereafter, as the program progresses through Milestone B and beyond, more detailed requirements and specifications will be developed, all linked and traceable to the KPPs. The government or its designated integrator needs to maintain control of the top-level requirements and exercise a heavy bias in the direction of minimizing changes before IOC, even if it means deferring great new ideas that can be put in the initial deployment; this needs to be done because the cost of such changes can be hidden and unappreciated even by the contractors who are asked to provide estimates for them. In its oversight role, the government should also ensure that the contractors are controlling lower-level, derived requirements, so that only requirements and requirements changes needed to meet the top-level end objectives are considered prior to IOC.

Intersystem and Intersegment Interface Management

Another key role for the systems engineers after Milestone B is to ensure that the interfaces among the segments and among the system and its users are clearly defined early in the design phase. A consequence of good architecture that partitions a system into separately procurable segments is that the interfaces among them need to be defined early and precisely, as they become key design drivers. There will always be a tendency for the segment designers themselves to want to define these interfaces late in the design phase when their designs are complete and the needed interface support is obvious. The problem with this approach is that it will invariably require the designers of interfacing segments to redesign with adverse schedule and cost impacts. The development of standards for data passed across interfaces, including data protocols, units of measure, coordinate systems, and so on, can be valuable in managing interfaces. The establishment of communities of interest across the interface boundaries for specific issues, such as data semantics, can also be useful.

Sharing Best Practices with Other Agencies

This study was requested by the Air Force; however, the committee's conclusions and recommendations apply to any acquisition program within the entire Department of Defense, regardless of the military department or service affiliation. Likewise, there are major acquisition and systems development programs in other government agencies that could also benefit from the recommendations of this study. For instance, the National Aeronautics and Space Administration (NASA) has embarked on the development of numerous new programs to accomplish the goals of the new U.S. space exploration policy to return to the Moon and travel to Mars and beyond. This new policy requires new space-launch systems, a replacement for the space shuttle, robotic technologies, and so on. Many of the required developments will be systems whose complexity will rival, if not exceed, the complexities and challenges of even the most critical and largest national security defense programs. NASA's companion to the DOD acquisition process has elements similar to those described in Chapter 1. A good systems engineering process that includes up-front development planning is just as beneficial to NASA as it is to DOD.

The Federal Aviation Administration (FAA) also faces the development of major new systems in the future to meet the goals of the Next Generation Air Transportation System.[9] This program requires the improvement of every node of the air transportation system (from the curbside origination of a traveler to the curbside destination of that traveler) to meet the demands of a future air system where superlarge airliners and very small air-taxis may mix with unmanned aerial vehicles. Again, the complexities of these developments beg for a good systems engineering process with good development planning.

In the cases of both NASA and the FAA, these developments will not occur in isolation within the specific agency. Instead, many of these efforts will be made with the cooperation of the DOD. Thus, cooperation in sharing good processes will not be a luxury. It will be a necessity.

[9]Federal Aviation Administration, 2007, *Online Fact Sheet: Next Generation Air Transportation System 2006 Progress Report*, March 14. Available at http://www.faa.gov/news/fact_sheets/news_story.cfm?newsId=8336. Last accessed on May 17, 2007.

PRE-MILESTONE A/B CHECKLIST

A summary of some of the most important outputs of the systems engineering process is presented here in the form of a checklist. Box 4-1 lists the outputs that a program or systems engineering manager might use in evaluating the completeness of the effort in each of the systems engineering functional areas described earlier. This checklist applies to Milestones A and B. Often a checklist item needed in preliminary form at Milestone A must be completed to a higher level of maturity at Milestone B. The committee notes in Chapter 2 that it believes the Air Force should include Milestone A in more of its development programs.

PREVIOUS RELEVANT STUDIES, FINDINGS, AND RECOMMENDATIONS

Several previous studies have commented on various aspects of the application of systems engineering to government acquisition programs. Their key findings and recommendations, which are in good agreement with those of this committee, are summarized below.

National Research Council Report

In 1993, the National Research Council's (NRC's) Air Force Studies Board (AFSB) completed a study entitled *An Examination of the Air Force's Pre-Milestone One Planning/Decision Process.*[10] In essence, that study reached conclusions similar to those of the General Accounting Office (now the Government Accountability Office; GAO) study discussed below: namely, that the critical leverage point in any program is matching the customer's needs with the developer's resources. The AFSB study concluded: "Where a strong partnership exists (e.g., as between the Air Force Space Command [AFSPACECOM] and the AFMC) the subsequent transition from user to developer is not a problem" (p. 2). This statement implies that there must be tight collaboration between user and developer in all pre-Milestone A activities, especially in all systems engineering activities.

Defense Science Board and Air Force Scientific Advisory Board Report (Young Panel Report)

In August 2002, the Under Secretary of Defense for Acquisition, Technology and Logistics; the Secretary of the Air Force; and the Under Secretary of the Air Force/Director of the National Reconnaissance Office chartered the Defense Science Board/Air Force Scientific Advisory Board Joint Task Force on Acquisition

---

[10]National Research Council, 1993, *An Examination of the Air Force's Pre-Milestone One Planning/Decision Process*, Washington, D.C.: National Academy Press.

BOX 4-1
Pre-Milestone A/B Checklist

Concept Development

1. Have at least two alternative concepts to meet the need been evaluated?

   The purpose of alternatives is to stimulate thinking to find the simplest, fastest, and cheapest solution.

2. Can an initial capability be achieved within the time that the key program leaders are expected to remain engaged in their current jobs (normally less than 5 years or so after Milestone B)? If this is not possible for a complex major development program, can critical subsystems, or at least a key subset of them, be demonstrated within that time frame?

   Achieving capabilities or demonstrating critical subsystems while key program leaders remain engaged is important to get the capability into service quickly and cost-effectively and to begin the process of incremental improvements based on operational experience.

3. Will risky new technology have been matured before Milestone B? If not, is there an adequate risk mitigation plan?

   The development of risky new technology in parallel with a major development program can be costly in terms of both time and money.

4. Have external interface complexities (including dependencies on other programs) been identified and minimized? Is there a plan to mitigate their risks?

   Complex, ill-defined, external requirements and interfaces can be a major source of requirements instability during the development phase. This can be of particular importance when a system must operate in a system-of-systems environment.

Key Performance Parameters and CONOPS

5. At Milestone A, have the KPPs been identified in clear, comprehensive, concise terms that are understandable to the users of the system?

   It is important that KPPs be expressed in terms understandable to all of the stakeholders. Failure to define the system's KPPs simply and clearly at Milestone A is a first step to requirements instability and overruns later.

continued

BOX 4-1 Continued

6. At Milestone B, are the major system-level requirements (including all KPPs) defined sufficiently to provide a stable basis for the development through IOC?

   Beginning development without a complete list of stable requirements is one of the key "seeds of failure" described in Chapter 4 in this report. It is important to complete requirements trade-offs prior to the development phase.

7. Has a CONOPS been developed showing that the system can be operated to handle the expected throughput and meet response time requirements?

   It can be costly to discover too late that the system as designed cannot be operated to meet its requirements.

Cost and Schedule Scoping

8. Are the major known cost and schedule drivers and risks explicitly identified, and is there a plan to track and reduce uncertainty?

   Identifying the major cost and schedule risk areas, with particular attention to this checklist and the six seeds of failure—inexperienced leadership, external interface complexity, system complexity, incomplete requirements at Milestone B, immature technology, and high reliance on new software—can help focus management on these issues early.

9. Has the cost confidence level been accepted by the stakeholders for the program?

   It is important that stakeholders understand the degree of risk so that the stakeholders will not disrupt the program as inevitable development program surprises unfold later on. It will generally not be possible by Milestone A or Milestone B to identify all the risk areas that might surface later in a development program, but a frank, early disclosure of known potentials for risk can help sustain stakeholder support later on.

Performance Assessment

10. Is there a sufficient collection of models and an appropriate simulation environment to validate the selected concept and the CONOPS against the KPPs?

continued

BOX 4-1 Continued

In large, complex programs, the development of models early on can be very important to later management of requirements changes and performance verification.

11. At Milestone B, do the requirements take into account likely future mission growth over the program life cycle?

    The committee advocates freezing new requirements and new technology insertion after Milestone B but also notes that making provisions in the initial requirements to facilitate later upgrades could have great long-term value.

Architecture Development

12. Has the system been partitioned to define segments that can be independently developed and tested to the greatest degree possible?

    Effective partitioning of a complex system can greatly reduce its development cost.

13. By Milestone A, is there a plan to have information exchange protocols established for the whole system and its segments by Milestone B?

    Such a plan developed early on can greatly reduce interface problems later in the development phase when they would be more difficult and costly to fix.

14. At Milestone B, has the government structured the program plan to ensure that the contractor addresses the decomposition of requirements to hardware and software elements sufficiently early in the development program?

    The histories of programs with cost and schedule overruns are replete with examples of large software developments that had to be redone because requirements from the hardware side were assigned or determined late.

Risk Assessment

15. Have the key risk drivers (not only the technology drivers) been identified?

    Identifying and managing risk early can pay large dividends; it is important to focus on the six "seeds of failure" (see item 8 above).

continued

BOX 4-1 Continued

Program Implementation Strategy

16. Does the government have access over the life of the program to the talent required to manage the program? Does it have a strategy over the life of the program for using the best people available in the government, the FFRDCs, and the professional service industry?

    Seasoned management is critical; the government's job is to find the best!

17. At Milestone A, is there a plan defining how the pre-Milestone B activity will be done, and by whom?

    Identifying the program and system managers early, identifying the FFRDC or SETA support needed, thinking through the use of competitive system concept contracts—all can have a decisive impact on the government's ability to select the best concept, to define by Milestone B system requirements that can remain stable through IOC, and to select the best development contractors.

18. Is there a top-level plan for how the total system will be integrated and tested?

    A well-thought-out strategy for verifying system performance, including optimum phasing of verification tests throughout the assembly process, and well-thought-out use of analytical models and external simulators can have a large positive impact on ultimate cost, schedule, and performance.

19. At Milestone B, have sufficiently talented and experienced program and systems engineering managers been identified? Have they been empowered to tailor processes and to enforce requirements stability from Milestone B through IOC?

    Seasoned leaders in these areas are critical to maintaining focus and discipline through IOC.

20. Has the government attempted to align the duration of the program manager's assignment with key deliverables and milestones in the program?

    A combination of assignment extension and time-certain milestones will help align incentives.

NOTE: KPP, key performance parameter; CONOPS, concept of operations; IOC, initial operational capability; FFRDC, federally funded research and development center; SETA, systems engineering and technical assistance.

of National Security Space Programs to review the acquisition of national security space programs, identify and characterize systemic problems, and recommend improvements. Four key points were made in the Young panel report (the task force's final report, frequently referred to as the "Young panel report" because the joint task force was chaired by A. Thomas Young):[11]

- Cost has replaced mission success as the primary driver in managing acquisition processes, resulting in excessive technical and schedule risk. We must reverse this trend and reestablish mission success as the overarching principle for program acquisition.
- The space acquisition system is strongly biased to produce unrealistically low cost estimates throughout the acquisition process. These estimates lead to unrealistic budgets and unexecutable programs.
- Government capabilities to lead and manage the acquisition process have seriously eroded. On this count, we strongly recommend that the government address acquisition staffing, reporting integrity, systems engineering capabilities, and program manager authority.
- While the space industrial base is adequate to support current programs, long-term concerns exist.

The chairs of the Defense Science Board (DSB) and Air Force Scientific Advisory Board (AFSAB) emphasized an overall government underappreciation for the importance of appropriately staffed and trained systems engineers for managing the technologically demanding and unique aspects of space programs.

In the Young panel report, the top five issues related to systems engineering were deemed to be the following:

- Lack of awareness of the importance, value, timing, accountability, and organizational structure of systems engineering for programs;
- General lack of availability within government and industry of adequate, qualified systems engineering resources for allocation to major programs;
- Insufficient systems engineering tools and environments to effectively execute systems engineering on programs;
- Lack of consistent and effective application of requirements definition, development, and management; and
- Poor initial program formulation.

---

[11]Defense Science Board/Air Force Scientific Advisory Board Joint Task Force, 2003, *Acquisition of National Security Space Programs*, Washington, D.C.: OUSD (AT&L). Available at http://www.acq.osd.mil/dsb/reports/space.pdf. Last accessed on April 2, 2007.

National Defense Industrial Association Report

The summary recommendations of the recent report by the National Defense Industrial Association (NDIA) were as follows:[12]

- Ensure institutionalization of effective systems engineering practices into program planning and execution.
- Integrate engineering planning within the acquisition life cycle to ensure adequate time and effort for SE early in the program life cycle.
- Emphasize the application of systems engineering practices and resources to the capability definition process to address warfighter needs and translation into executable programs.
- Grow systems engineering expertise through training, career incentives, and broadening "systems thinking" into other disciplines.
- Strengthen and clarify policy and guidance regarding the use of collaborative environments, models, simulations, and other automated tools.

Government Accountability Office Reports

Pertinent to this committee's discussion of pre-Milestone A SE, the GAO, in its 2003 report *Defense Acquisitions: Improvements Needed in Space Systems Acquisition Management Policy,*[13] stated that the most leveraged decision point in a program is matching the customer's needs with the developer's resources. This initial decision sets the stage for the eventual outcome. In successful programs, negotiations and trade-offs occur before product development is started. As an example, GAO pointed out that at the time the decision was made to accelerate the Advanced Extremely High Frequency (AEHF) Program (see the MILSATCOM case summary in Chapter 2), DOD had neither the funding nor the manpower to accomplish the task.

The GAO also reported that a primary problem affecting Air Force space programs is that often the users refuse to relax rigid requirements to more closely match technical capabilities that are achievable. For instance, for one element of the Space Based Infrared Systems (SBIRS) Program (see the SBIRS case study in Chapter 2), it became apparent that a lack of knowledge of program challenges led to overly optimistic schedules and budgets. Attempts to stay on schedule by

[12]National Defense Industrial Association, 2006, *Top Five Systems Engineering Issues Within Department of Defense and Defense Industry*. Available at http://www.ndia.org/Content/ContentGroups/Divisions1/Systems_Engineering/Top_5_Systems_Engineering_Issues.pdf. Last accessed on April 2, 2007.

[13]Government Accountability Office, 2003, *Defense Acquisitions: Improvements Needed in Space Systems Acquisition Management Policy*, September. Available at http://www.gao.gov/new.items/d031073.pdf. Last accessed April 2, 2007.

approving critical milestones without meeting critical criteria led to higher costs and even further schedule erosion.[14]

The 2005 GAO study *Space Acquisitions: Stronger Development Practices and Investment Planning Needed to Address Continuing Problems*[15] was done at the request of Congress for input on problems relating to DOD's space systems acquisition. To meet this request, the GAO drew on its previous reports related to the causes of acquisition problems, underlying incentives and pressures, and potential solutions. The testimony resulting from the study partially repeated conclusions from the 2003 GAO report listed above. It concluded that programs typically do not achieve a match between requirements and resources at program start. Specifically:

- Either requirements are not adequately defined early, or they are changed dramatically once the program is underway.
- Technologies are typically not mature enough to be included in product development.
- There are deficiencies in the space acquisition workforce, contractor capabilities, and funding available for testing of space technologies.
- There is a tendency for programs to take on technology development that should occur in the science and technology (S&T) community.
- DOD starts more programs than it can afford in the long run, forcing programs to underestimate cost and overpromise capability.
- The most pertinent of GAO's recommendations was to employ the techniques of SE to close gaps between available technologies and customer needs before committing to new product development. Also, GAO recommended that DOD develop plans for addressing the shortage of staff with science and engineering backgrounds.[16]

Defense Acquisition Performance Assessment

The major findings of the 2006 *Defense Acquisition Performance Assessment* (DAPA) study[17] were these:

- Strategic technology exploitation is a key U.S. advantage. Opportunities need to be identified early.

---

[14] Ibid.

[15] Government Accountability Office, 2005, *Space Acquisitions: Stronger Development Practices and Investment Planning Needed to Address Continuing Problems*, July 12. Available at http://www.gao.gov/new.items/d05891t.pdf. Last accessed April 2, 2007.

[16] Ibid.

[17] Ronald Kadish, Gerald Abbott, Frank Cappuccio, Richard Hawley, Paul Kern, and Donald Kozlowski, 2006, *Defense Acquisition Performance Assessment*. Available at http://www.acq.osd.mil/dapaproject/documents/DAPA-Report-web/DAPA-Report-web-feb21.pdf. Last accessed on April 2, 2007.

- The U.S. economic and security environments have changed; for example, there are fewer prime contractors, smaller production runs, reduced plant capacity, fewer programs, unpredictable threats.
- The acquisition system must deal with instability of external funding.
- The DOD management model is based on a lack of trust. Quantity of review has replaced quality. There is no clear line of responsibility, authority, or accountability.
- Oversight is preferred to accountability.
- Oversight is complex, not process- or program-focused (as it should be).
- The complexity of the acquisition process increases costs and draws out the schedule.
- Incremental improvement applied solely to the "little a" acquisition process[18] requires all processes to be stable—but they are not.

The DAPA report dealt with larger issues of acquisition culture and policy. A member of this committee, who was the chair of the DAPA study, emphasized that a successful response to the instabilities caused by the current process or proper program initiation as envisioned requires early and detailed SE practices.

FINDINGS AND RECOMMENDATIONS

The considerations discussed in this chapter led the committee to develop the following findings and recommendations:

Finding 4-1. Attention to a few critical systems engineering processes and functions particularly during preparation for Milestones A and B is essential to ensuring that Air Force acquisition programs deliver products on time and on budget.

Today's weapons systems provide unprecedented capabilities but also involve complex interfaces with external command, control, and communications systems and rely on a greater volume of software than ever before. Early decisions on the weapons system requirements and capabilities have a disproportionately large impact on program cost and schedule. The committee also recognizes that a lack of flexibility (a result of overly rigid processes or a lack of trust among program participants or stakeholders) can limit the ability of a program manager to change early decisions that warrant changing.

The committee found many gaps and inconsistencies in the way that the Air Force manages pre-Milestone A activities. The committee heard from presenters of some cases in which required documents were completed pro forma and filed away, never to be seen again, or for which required steps were skipped

[18]The Acquisition—"Big A"—system is often believed to be a simple construct that efficiently integrates three independent processes: requirements, budgeting, and acquisition. "Little a," on the other hand, refers to the acquisition process that focuses on "how to buy" in an effort to balance cost, schedule, and performance; it does not include requirements and budgeting.

completely. The current practice of initiating programs at Milestone B denies the acquisition review authority the earlier opportunity (at Milestone A) to make judgments about the maturity of the technologies on which the program is based and to decide whether technologies need to be further developed prior to making a Milestone B commitment to system development and demonstration.

Recommendation 4-1. The Air Force leadership should require that Milestones A and B be treated as critical milestones in every acquisition program and that a checklist such as the "Pre-Milestone A/B Checklist" suggested by the committee (see Box 4-1 in this chapter) be used to judge successful completion.

A rigorous, standard checklist of systems engineering issues should be addressed by each program through both the pre-Milestone A and pre-Milestone B phases. The committee's recommended 20-item checklist is shown in Box 4-1. While the committee considers that each item on the checklist is important, it calls attention to several items that warrant further discussion:

- Checklist item 2 recognizes that the world changes too fast to be friendly to long development cycles. The committee believes that the Air Force should strive to structure major development programs so that initial deployment is achieved within, say, 3 to 7 years. Thirty years ago, this was a typical accomplishment—for example, nearly 40 years ago, the Apollo program put the first man on the Moon in fewer than 8 years.
- The development time issue is addressable by applying systems engineering to items 3, 4, and 13 through 15 before Milestones A and B. The definition of clear KPPs by Milestone A and clear requirements by Milestone B that can remain stable through IOC can be essential to an efficient development phase. It is also important that critical technologies be sufficiently mature prior to starting SDD. The committee observed that although today's systems are not necessarily more complex internally than those of 30 years ago, their external complexity often is greater, because today's systems are more likely to try to meet many diverse and sometimes contradictory requirements from multiple users. This kind of complexity can often lead to requirements being changed between Milestone B and IOC, and it can lead to relying on immature technology.
- Item 19 of the checklist stresses the importance of placing experienced, domain-knowledgeable managers in key program positions. The committee has observed that many of the truly extraordinary development programs of the past, such as Apollo, the Manhattan Project, the early imaging satellite programs, the U-2, the fleet ballistic missile system, and nuclear submarines, were managed by relatively small (and often immature) agencies with few established processes and controls. In that environment, dedicated managers driven by urgent missions accomplished feats that often seem incredible today. The committee believes that the

> accumulation of processes and controls over the years—well meant, of course—has stifled domain-based judgment that is necessary for timely success. Formal SE processes should be tailored to the application. But they cannot replace domain expertise. In connection with Item 19, the committee recommends that the Air Force place great emphasis on putting seasoned, domain-knowledgeable personnel in key positions—particularly the program manager, the chief system engineer, and the person in charge of "requirements"—and then empower them to tailor standardized processes and procedures as they feel is necessary.

One key pre-Milestone A task is the analysis of alternatives (AoA), which entails evaluating alternative concepts and comparing them in terms of capabilities, costs, risks, and so on. Checklist items 1 through 4, 12, and 13 should be completed before the AoA, while items 5 through 11 and 14 through 20 may be addressed after the AoA.

Finding 4-2. The Air Force used to have a development planning organization that applied pre-Milestone A systems engineering processes to a number of successful programs, but that organization was allowed to lapse.

The role of the Air Force development planning organization, which was within the Air Force Systems Command, was to provide standard evaluation tools and perform pre-Milestone A systems engineering functions across acquisition programs. The early 1990s saw an erosion of this front-end planning organization along with its funding as the Air Force Systems Command (now the Air Force Materiel Command [AFMC]) began to play a decreasing role in program execution.

In the opinion of several speakers who met with the committee, one main reason for the erosion of funding was a lack of congressional support for the planning function. The specific budget "program element" (PE) for the Air Force planning function was PE 65808. This PE funded a robust planning process in the Air Force for many years. Again, several programs noted in Chapter 2 had their roots in the pre-Milestone A analysis, concept development, and prototyping funded by PE 65808 in the past. The funding for PE 65808 began to decline in the early 1990s as Congress started to reduce the funds appropriated. The rationale for this decline is not documented. However, one committee member recalls specific meetings with key congressional staffers who were responsible for approving the funding of PE 65808. At one of these meetings a staffer stated that this PE ultimately leads to new programs that Congress will have to fund in the future, and so if Congress does not fund PE 65808, then the Air Force will not develop new unaffordable systems.

Recommendation 4-2. A development planning function should be established in the military departments to coordinate the concept development and refinement phase of all acquisition programs to ensure that the capabilities required by the country as a whole are considered and that unifying strategies such as network-centric operations and interoperability are addressed.

The Air Force and the other military services should establish a development planning organization like that which existed in the early 1990s.

The roles and functions of the various organizations involved in acquiring major weapons systems need to be clearly defined. The responsibility for executing systems engineering and program management in the pre-Milestone A and B phases should be vested in the military departments that do the actual development planning functions. This should not be the responsibility of the Office of the Secretary of Defense (OSD) or of the Joint Staff. Instead, those offices need to enable the creation and functioning of military department development planning organizations with policy measures, and, where appropriate, resources. The Joint Staff, under the auspices of the Joint Requirements Oversight Council (JROC), may help to define the requirements for major programs in the course of the development planning process, but it should not run the process itself.

The existence of "joint" programs or a program such as Missile Defense, which has several related systems being developed by different military services, requires clear guidance from both OSD and the Joint Staff about who is in charge. These programs need to be harmonized and integrated by the responsible integrating agency. However, development planning activities should still take place in the military departments where the expertise resides. Consequently, the development planning should be managed by that agency.

While this committee cannot predict how Congress will view the revival of a good planning process to support pre-Milestone A program efforts, it is still important for the Air Force and DOD to make the case for the critical importance of this process before Congress and others. A development planning process is important not to start new programs, but rather to ensure that any new program (or a new start of any kind) is initiated with the foundation needed for success. Funding for this planning function needs to be determined by the military services, including both the acquisition communities and those (the warfighters) who generate the operational requirements.

CONCLUDING THOUGHTS

Many of the conclusions reached and recommendations made by the committee are similar to those of previous reviews. Most of the past recommendations were never implemented, so one of this committee's most critical thoughts relates to the importance of implementation. Successful implementation of these recommendations requires the "zipper concept"—making connections at all levels, from the senior leadership of the Air Force and DOD down to the working levels within key program management offices and supervisory staffs.

Appendixes

Appendix A

Biographical Sketches of Committee Members

Paul G. Kaminski, *Chair*, a member of the National Academy of Engineering, is chairman of the board and chief executive officer of Technovation, Inc., a consulting company dedicated to fostering innovation and the development and application of advanced technology. Dr. Kaminski is also a senior partner in Global Technology Partners, LLC. He is a former Under Secretary of Defense for Acquisition and Technology, and served as an Air Force officer, directing the Low Observable ("Stealth") Program and the development of advanced National Reconnaissance Space Systems. His professional activities include serving on the National Reconnaissance Office (NRO) Technical Advisory Board, the FBI Director's Advisory Board, and the Senate Select Committee on Intelligence Technical Advisory Board. He is a fellow of the Institute of Electrical and Electronics Engineers as well as of the American Institute of Aeronautics and Astronautics, and is a director of the Atlantic Council. He has authored numerous publications dealing with inertial and terminal guidance system performance, simulation techniques, Kalman filtering, and numerical techniques applied to estimation problems. Dr. Kaminski received a Ph.D. in aeronautics and astronautics from Stanford University, M.S. degrees in aeronautics and astronautics and in electrical engineering from the Massachusetts Institute of Technology (MIT), and a B.S. from the Air Force Academy.

Lester L. Lyles, *Vice Chair*, retired from the Air Force Materiel Command, Wright-Patterson Air Force Base, Ohio, as commander. He entered the Air Force in 1968 as a distinguished graduate of the Air Force Reserve Officer Training Corps Program. During his career in the United States Air Force, he has served in various assignments, including those as program element monitor of the short-range attack missile at USAF Headquarters, special assistant and aide-de-camp to the commander of Air Force Systems Command, Avionics Division chief in the F-16 Systems Program Office, director of Tactical Aircraft Systems at Air Force Systems Command (AFSC) Headquarters, and as director of the Medium-Launch Vehicles program and Space-Launch Systems offices. General Lyles became AFSC Headquarters assistant deputy chief of staff for requirements in 1989 and deputy chief of staff for requirements in 1990. In 1992, he became vice commander of Ogden Air Logistics Center, Hill Air Force Base, Utah. He served as commander of the center until 1994, after which he was assigned to command the Space and Missile Systems Center at Los Angeles Air Force Base, California, until 1996. General Lyles became the director of the Ballistic Missile Defense Organization in 1996. In May 1999, he was assigned as vice chief of staff at USAF Headquarters. General Lyles received an M.S. degree in mechanical and nuclear engineering from New Mexico State University and a B.S. in mechanical engineering from Howard University.

Dev A. Banerjee serves as senior director and functional leader of systems engineering in the Integrated Defense Systems organization of the Boeing Company. As director of systems engineering at IDS, he has functional responsibility for the systems engineering disciplines of systems engineering measurement and control; affordability; system modeling and simulation; system integration, verification and validation; human system integration; operations/systems analysis; reliability, maintainability, and systems health; systems safety; systems security; systems architecture and definition; and certification/qualification. Dr. Banerjee chairs the Systems Engineering/Concept Definition Sub-Council in integrating engineering resources in systems engineering and flight engineering disciplines across Boeing. He serves as the IDS engineering focal point for the Boeing Intellectual Property Council and as the IDS engineering focal point for managing an integrated engineering and supplier management plan for improved supplier performance. Dr. Banerjee holds a D.Sc. from Washington University in St. Louis in mechanical/aerospace engineering.

Thomas W. Blakely is vice president of engineering for Lockheed Martin Aeronautics Company. Since his appointment in June 2003, Mr. Blakely has led the engineering organization, which includes almost 8,700 engineers, scientists, and technicians engaged in delivering technical solutions for high-performance military aircraft and systems. Mr. Blakely graduated from Texas A&M University in 1979 with a B.S. degree in aerospace engineering and directly joined the former Lockheed-California Company in Burbank. He was assigned to the Maritime Patrol and Anti-submarine Warfare business group and was involved with a variety of development programs related to the P-3 Orion and CP-140 Aurora aircraft. In 1984, he was assigned to Lockheed's office in Arlington, Va., to represent the Patrol Aircraft Engineering Division at the Naval Air Systems Command. He returned to Burbank, Calif., in 1986 and was promoted to engineering program manager for P-3C Orion programs. In 1988, he was reassigned to lead the preliminary design and EMD team responsible for the development of aircraft subsystems for the P-7 maritime patrol aircraft. In 1990, he returned to the Washington, D.C., area for a second tour in the company's Arlington, Va., office, again working with the Maritime Patrol Engineering office at the Naval Air Systems Command. In 1991, Mr. Blakely transferred to Lockheed Aeronautics in Marietta, Ga., assuming responsibility for all of the company's International Maritime Patrol Aircraft Engineering programs. In 1996, he was selected to lead the C-130J systems verification and flight test team and played a significant technical leadership role in civil certification of the new 382J and development and testing of the C-130J military configuration. He was subsequently promoted to chief systems engineer and, ultimately, chief engineer for C-130 programs. In August 2000, Mr. Blakely was promoted to the position of vice president as the deputy for engineering for Lockheed Martin Aeronautics Company. In this role, he was involved with the consolidation of engineering operations, personnel, processes, and tools at the new company's three sites into a single organization. The next few years also included several special technical leadership assignments on both the C-130J Program and the C-5 Avionics Modernization Program. He was selected as Lockheed Martin Aeronautics Company vice president for engineering in the spring of 2003. In January 2004, he joined the Joint Strike Fighter (JSF) executive leadership team on special assignment as the technical director for the JSF Development Program. While he was on this assignment, the program plan and strategy were restructured, and the preliminary design configuration and arrangement of the aircraft went through a substantial design iteration to reduce weight and improve operational suitability. Mr. Blakely received a B.S. in aerospace engineering from Texas A&M University.

Natalie W. Crawford, a member of the National Academy of Engineering, is a senior fellow at the RAND Corporation. Immediately prior to this position, from 1997 to 2006, she held the position of vice president of the RAND Corporation and director of Project AIR FORCE (PAF). It was her responsibility to ensure that the research agenda of PAF addressed problems of greatest enduring importance to the Air Force, and that the research was of the highest possible quality and responsiveness. She has worked at the RAND Corporation for more than 40 years and has deep, substantive technical and operational knowledge and experience in areas such as conventional weapons, attack and surveillance avionics, fighter and bomber aircraft performance, aircraft survivability, electronic combat, theater missile defense, force modernization, space systems and capabilities, and nonkinetic operations. She has been a member of the Air Force Scientific Advisory Board since 1988, and served as its vice chair in 1990 and its co-chair from 1996 to 1999. She has served on numerous advisory committees. Mrs. Crawford has a B.S. in mathematics from the University of California at Los Angeles where she also pursued graduate study in applied mathematics and engineering.

Stephen E. Cross is a vice president of the Georgia Institute of Technology and the director of the Georgia Tech Research Institute. He also holds faculty appointments as a professor in industrial and systems engineering and as a professor in computer science. Before joining Georgia Tech in 2003, he was the director and chief executive officer of the Software Engineering Institute, a Department of Defense-sponsored federally funded research and development center at Carnegie Mellon University. Dr. Cross was a member of the Defense Science Board Task Force on Defense Software in 2000. He currently serves on the Air Force Scientific Advisory Board and the Defense Advanced Research Projects Agency (DARPA) Panel for Information Science and Technology. Dr. Cross is a fellow of the Institute of Electrical and Electronics Engineers. A retired Air Force officer, he attended the Air Force Test Pilot School (Flight Test Engineer Course) and served in various research and development assignments as a software engineer for the F-16, F-15, and B-1A programs; a flight test engineer in the Air Launched Cruise Missile program; an assistant professor at the Air Force Institute of Technology; a research manager at Air Force Wright Aeronautical Laboratories; and a program manager at DARPA. Dr. Cross received his Ph.D. from the University of Illinois at Urbana-Champaign, his M.S. in electrical engineering from the Air Force Institute of Technology, and his B.S. in electrical engineering from the University of Cincinnati.

Gilbert F. Decker is a private consultant for several clients, including the Boeing Corporation, the United States Navy, and Walt Disney Imagineering, where he was previously the executive vice president of engineering and production. He has also served as a commissioned officer in the U.S. Army and as a colonel in the U.S. Army Reserve. Before becoming a private consultant, he held several distinguished positions, including those of president and chief executive officer of the Penn Central Federal Systems Company, president and chief executive officer of Acurex Corporation, and Assistant Secretary of the Army/Research, Development, and Acquisition. Mr. Decker currently serves on the National Advisory Council for the Johns Hopkins University, Whiting School of Engineering, and on the Board on Army Science and Technology of the National Research Council. He acts as the director of Alliant TechSystems, Anteon Corporation, and the Allied Research Corporation. Mr. Decker is also a trustee for the Hertz Foundation and for the Association of the U.S. Army. He received an M.S. degree in operations research from Stanford University and a B.S. degree in engineering science and electrical engineering from the Johns Hopkins University.

Llewellyn S. Dougherty is the vice president, Special Programs, for Raytheon Company. He has served in other areas of the company, including sensors and communications, radar systems, and reconnaissance systems. Prior to his career at Raytheon, he was technical assistant to the director of the Defense Advanced Research Projects Agency. His areas of expertise include avionics, digital computers, software, systems engineering, and systems safety. Dr. Dougherty received a Ph.D. in digital systems engineering from the Air Force Institute of Technology, an M.S. in aeronautics and astronautics from the Massachusetts Institute of Technology (MIT), and a B.S. in astronautics and engineering sciences from the U.S. Air Force Academy.

John V. Farr is currently a professor and associate dean for academics in the School of Systems and Enterprises for the Stevens Institute of Technology. He is also the founder and principal of Farr Engineering and Management Consulting in Cornwall-on-Hudson, N.Y., where he provides quantitative business and decision-support solutions to a wide variety of industrial clients. Dr. Farr started his technical career at U.S. Army Engineers Waterways Experiment Station. He joined the faculty of the U.S. Military Academy at West Point in 1992 as the first permanent engineering professor. He was appointed to the rank of full professor at West Point in 2000. He joined the faculty at Stevens Institute of Technology in the fall of 2000 as a professor and the founding director of the Department of Systems Engineering and Engineering Management. He is a fellow in the American Society of Civil Engineers and the American Society of Engineering Management (ASEM) and has authored more than 100 technical publications. He also serves on the Army Science Board and is the past president of ASEM. Dr. Farr received a Ph.D. from the University of Michigan, an M.S. from Purdue University, and a B.S. from Mississippi State University in civil engineering.

James H. Frey is currently principal at Frey Associates. Prior to this position, he served as senior vice president, group executive for the Litton Information Systems Group until Northrop Grumman's purchase of Litton in April 2001. In this position, he led the $1.4 billion group of Litton businesses involved in information technology services. Mr. Frey joined Litton in 1988 as president of Litton Itek Optical Systems and held that position until 1996 when he was appointed vice president of strategic development. Three years later, he became the information systems group executive. He was appointed president of TASC in 1999, a company specializing in systems engineering and program management for large space, intelligence, and information management systems. Prior to joining Litton, Mr. Frey spent many years at General Electric, where he rose to the position of general manager of the Spacecraft Division. Since retiring from his position as president of Northrop Grumman TASC in March 2002, he has served on the Advisory Board for the National-Geospatial Intelligence Agency and on special panels supporting the management of the National Reconnaissance Office and the Director of Central Intelligence. He serves on the board of directors of Nortel Government Solutions and Electronic Sensor Technologies and advises a number of private-sector clients, including Next Century, Appistry, Northrop Grumman, and Boeing. Mr. Frey received his B.S. in electrical engineering from Duke University.

Robert A. Fuhrman, a member of the National Academy of Engineering, is retired vice chairman of the board, president, and chief operating officer of the Lockheed Corporation, and a former chair of the Air Force Science and Technology Board (now the Air Force Studies Board). Mr. Fuhrman has had a distinguished career, having served as Lockheed's president and chief operating officer and group president for missiles and space, as well as in numerous other positions. He was directly responsible for the systems engineering of the early fleet ballistic missile (Polaris) programs. He received an M.S. in fluid mechanics and dynamics from the University of Maryland and a B.S. in aeronautical engineering from the University of Michigan.

David J. Gorney is vice president of space program operations at the Aerospace Corporation and is responsible for the company's support to all Air Force and Navy satellite programs. Before being named vice president, Dr. Gorney served as general manager of the Navigation Division, directing technical initiatives associated with key upgrades to the Defense Department's Global Positioning System. Dr. Gorney joined the Aerospace Corporation in 1979 as a mathematician in the Space Sciences Laboratory. Other positions that he held within the laboratory included those as a member of the technical staff, as a research scientist, and as a manager and a director. Subsequent to these assignments he served as principal director of four organizations: the Office of Research and Technical Applications, the Office of Research and Engineering, the Defense Support Program, and Meteorological Satellite Systems. He was corporate chief architect/engineer before being named general manager of the Navigation Division in October 2002. Dr. Gorney received Ph.D. and M.S. degrees in atmospheric sciences from the University of California, Los Angeles, and a B.S. degree in physics from the University of Bridgeport in Connecticut.

John M. Griffin is president of Griffin Consulting, providing systems engineering and program management services to large and mid-sized aerospace firms. He provides strategy planning initiatives for corporations, reviews ongoing programs to assess progress and recommend corrective actions, and participates with industry and government in developing program strategy and implementation tactics. During his civilian career with the Air Force, Mr. Griffin served in a diverse spectrum of capacities with a range of assignments and special duties. He served on numerous special panels, two of which formed the structure of the Air Force Materiel Command. Mr. Griffin was on the development team for groundbreaking technology revolutions in weapon systems, including stealth, unmanned vehicles, hypersonics, and cruise missiles. He retired from the Air Force in 1997.

Mr. Griffin holds an M.S. degree in electrical engineering from the Air Force Institute of Technology and a B.S. degree in aeronautical engineering from the University of Detroit.

Wesley L. Harris, a member of the National Academy of Engineering, is the Charles Stark Draper Professor and head of the Department of Aeronautics and Astronautics at the Massachusetts Institute of Technology. His research focuses on theoretical and experimental unsteady aerodynamics and aeroacoustics, on computational fluid dynamics, and on the impact of government policy on the procurement of high-technology systems. Prior to this position he served as the associate administrator for aeronautics at NASA. He has also served as the vice president and chief administrative officer of the University of Tennessee Space Institute. Dr. Harris received a Ph.D. and an M.S. in aerospace and mechanical sciences from Princeton University and a B.S. in aerospace engineering from the University of Virginia.

Ronald T. Kadish (U.S. Air Force, ret.) is a vice president and partner at Booz Allen Hamilton. He is active in the Defense Team's Aerospace Market Group and focuses his efforts in the Command, Control, and Communications Market Thrust Team. General Kadish joined Booz Allen Hamilton in February 2005 after retiring from the Air Force as a lieutenant general. He is a former Air Force pilot with more than 2,500 flying hours, and he held a variety of senior systems acquisition, program management, and command positions, including those as program director for the F-15, F-16, and C-17; director of the Missile Defense Agency; and commander of the Center of Excellence for Command and Control Systems. General Kadish received an M.B.A. from the University of Utah and a B.S. in chemistry from St. Joseph's University in Philadelphia, Pa.

Robert H. Latiff is vice president, chief engineer, and technology officer, Space and Geospatial Intelligence, Science Applications International Corporation (SAIC). Prior to joining SAIC, General Latiff was deputy director for systems engineering, National Reconnaissance Office (NRO). He retired from the Air Force as a major general in 2006. As the NRO's systems engineer, General Latiff managed the NRO acquisition process and was the functional manager for NRO-wide systems engineering. He worked with senior program managers to define the Integrated NRO Architecture for space-based reconnaissance and intelligence systems. While at the NRO, General Latiff also served as director, Advanced Systems and Technology. General Latiff received his commission after completing the Army Reserve Officer Training Corps program at the University of Notre Dame and subsequently transferred to the Air Force in 1980. He has served on the staffs of Headquarters U.S. Air Force and of the Secretary of the Air Force. In a previous assignment with the Air Force's Electronic Systems Center, he was the program director for the E-8C, Joint Surveillance Target Attack Radar

System. General Latiff then commanded the Joint U.S. and Canadian Cheyenne Mountain Operations Center in Colorado Springs, Colo. He received his Ph.D. and M.S. in materials science and a B.S. in physics from the University of Notre Dame. He is a member of the National Research Council's National Materials Advisory Board (NMAB).

Alden V. Munson, Jr., is the deputy director of national intelligence for acquisition in the Office of the Director of National Intelligence, a position he assumed during the term of this study. He was most recently a consultant in intelligence and defense to government and industry. Mr. Munson was senior vice president and group executive of the Litton Information Systems Group. Previously he served as a vice president at TRW in space and ground systems for command and control and intelligence programs. He also served as vice president, operations, in the TRW credit business. Previously, he served as program manager for numerous intelligence systems development projects and led major new business pursuits. Mr. Munson began his career at the Aerospace Corporation, where he provided system engineering and data system analysis and support to many space programs. He was a founding director of Paracel, Inc. (subsequently sold to Aplera) and has held board positions with bd Systems and the Armed Forces Communications and Electronics Association. He serves as an adviser to the San Jose State University College of Engineering. Mr. Munson received a master's degree in mechanical engineering from the University of California, Berkeley, and a B.S. degree in mechanical engineering with distinction and departmental honors from San Jose State University. He later completed extensive coursework in computer science at University of California at Los Angeles and attended executive programs at Harvard University (Competition and Strategy) and Stanford University (Management of High Technology Enterprises). In 1997, he was named a Distinguished Graduate of the San Jose State University College of Engineering, and in 2000, the National Reconnaissance Office named Mr. Munson a Pioneer of National Reconnaissance.

Mark K. Wilson, president, Mark Wilson Consulting, retired from the United States Air Force as director of the Center for Systems Engineering, Air Force Institute of Technology, Wright-Patterson Air Force Base, Ohio. He was responsible for planning, directing, and evaluating the development and sustainment efforts of all aspects of the Center for Systems Engineering. He has 38 years of systems engineering acquisition experience, including with capability-based systems engineering, and graduate education as well as extensive experience in flight systems, materials, low observables, and structural technology on the B-2 and F-15 programs. Mr. Wilson is an associate fellow of the American Institute of Aeronautics and Astronautics and a member of the corporate advisory board of the International Council on Systems Engineering and the National Defense

Industrial Association, Systems Engineering Division, Government Steering Group. Mr. Wilson holds M.S. degrees in management and management science from Stanford University and the University of Dayton, respectively, and a B.S. degree in aerospace engineering from Purdue University.

Appendix B

Meetings and Speakers

MEETING 1
JANUARY 8-9, 2007
THE KECK CENTER OF THE NATIONAL ACADEMIES
WASHINGTON, D.C.

Briefing to NRC Committee on Systems Engineering
Terry Jaggers, Deputy Assistant Secretary of the Air Force for Science, Technology, and Engineering

DOD Systems and Software Engineering
Mark Schaeffer, Office of the Under Secretary of Defense for Acquisition, Technology, and Logistics

Thoughts on Systems Engineering
Maj Gen Mark D. Shackelford, Director, Plans and Requirements Headquarters, Air Force Space Command

Air Force Acquisition: Transforming Acquisition
The Honorable Sue C. Payton, Assistant Secretary of the Air Force for Acquisition

Air Force Materiel Command Pre-Milestone A Systems Engineering
Winifred Okumura, Deputy Director, Intelligence and Requirements, HQ AFMC/A2/5

Applying Systems Engineering to Pre-Milestone A Activities
Col Jim Horejsi, Chief Engineer, Space and Missile Systems Center

Acquisition Transformation and Accelerating Change
The Honorable James I. Finley, Deputy Under Secretary of Defense for Acquisition, Technology, and Logistics

Early Planning Systems Engineering: ASC Implementation
Chris E. Leak, Chief, Capability Development Branch, Aeronautical Systems Center

The Counterspace Architecting Process
Roberta M. Ewart, Chief Scientist, Air Force Space and Missile Systems Center

An Assessment of the National Security Software Industrial Base
Piere Chao, Center for Strategic and International Studies

Systems Engineering Issues
Bob Rassa, Raytheon; Chair, NDIA Systems Engineering Division

National Reconnaissance Office
Doug Loverro, Associate Director, Imagery Systems Acquisition and Operations, National Reconnaissance Office

A Perspective on System Engineering: Delivering Capabilities
Dave Jacques, Curriculum Chair, Air Force Center for Systems Engineering

MEETING 2
JANUARY 31-FEBRUARY 1, 2007
THE ARNOLD AND MABEL BECKMAN CENTER
IRVINE, CALIFORNIA

Analytic Services (ANSER) Systems Engineering Emphasis
Ruth David, President and CEO, Analytic Services, Inc. (ANSER)

Space Program Acquisition: Systems Engineering and Programmatic Improvements
William F. Ballhaus, Jr., President and CEO, Aerospace Corporation

Space Based Infrared Systems Wing
Col Randall Weidenheimer, Space Based Infrared Systems Wing, Space and Missile Systems Center

An Assessment of the Continuum of the Systems Engineering Process: AFIT's Systems Engineering Case Studies
John Griffin, Committee Member

The Next Generation of Air Force Systems Engineering: Application to Capability Planning
Jeff Loren, Senior Acquisition Technical Manager, Systems Engineering Policy and Programs, Office of the Deputy Assistant Secretary of the Air Force for Science, Technology, and Engineering

System of Systems Engineering Challenges
BGen Ellen Pawlikowski, Commander, MILSATCOM Systems Wing, Space and Missile Systems Center

Developing Space-Based Capabilities: Has DOD Lost the Recipe?
Myron Hura, Senior Engineer, RAND Corporation

MEETING 3
FEBRUARY 27-28, 2007
THE KECK CENTER OF THE NATIONAL ACADEMIES
WASHINGTON, D.C.

C-5 Avionics Modernization Program (AMP): A Technology Insertion Program Case Study
John E. Weaver, Director, Air Systems Design and Integration, Lockheed Martin Aeronautics

Air Force Perspective on Planning, Requirements, and Systems Engineering
Harry Disbrow, Assistant Director of Operational Capability Requirements, Deputy Chief of Staff for Air, Space and Information Operations, Plans and Requirements, USAF/A5R

Program Review on the Future Combat Systems
Major General Charles A. Cartwright, Program Manager, Future Combat Systems (Brigade Combat Team)

The Value of Systems Engineering: Some Perspectives from Commercial Industry
Dinesh Verma, Associate Dean and Professor, Charles V. Shaefer, Jr., School of Engineering, Stevens Institute of Technology

Early Systems Engineering (SE) in Context of Acquisition Initiatives
Kristen Baldwin, Office of the Under Secretary of Defense (Acquisition, Technology, and Logistics)

Pre-Milestone A/B Engineering of Highly Capable, Complex, and Affordable Air Force Systems
Jim Mattice, Senior Consultant, Universal Technology Corporation

Defense Acquisition Performance Assessment: Report Summary Briefing
Lt. Gen. Ronald Kadish (USAF, ret.), Booz Allen Hamilton

MEETING 4
MARCH 27-28, 2007
THE KECK CENTER OF THE NATIONAL ACADEMIES
WASHINGTON, D.C.

Discussion with Committee
Brig. Gen. Janet Wolfenbarger, Director, Intelligence and Requirements Directorate, and Special Assistant for Command Transformation to the Commander, Headquarters, Air Force Materiel Command

Appendix C

What Is Systems Engineering?

DEFINITIONS

Before one can develop a definition for systems engineering (SE), one must develop a taxonomy for what constitutes a system, including the hierarchy of the elements of any complex system (Table C-1). Depending on one's perspective, the term "system" can be used to describe any element of this hierarchy, from a part to a system of systems (SoS).

SE has been defined in many different ways, ranging from high-level statements to detailed process overviews. These definitions are often tailored to illustrate a particular background or perspective. However, most accepted definitions have common themes describing a top-down process that is life-cycle-oriented and involves the integration of functions, activities, and organizations.[1] The most widely accepted definitions of systems engineering are presented in Table C-2.

To capture all of the essential elements addressed in this report, the committee chose a fairly detailed, three-part definition of SE:[2] (1) SE is the translation of a need or deficiency into a system architecture through the application of rigorous methods to the iterative process of functional analysis, allocation, implementation, optimization, test, and evaluation; (2) it is the incorporation of all technical parameters to ensure compatibility among physical and functional interfaces, and hardware and software interfaces, in a manner that optimizes system definition and design; (3) it is the integration of performance, manufacturing, reliability,

[1]Benjamin Blanchard and Wolter Fabrycky, 2005, *Systems Engineering and Analysis* (4th Edition), Englewood Cliffs, N.J.: Prentice Hall.

[2]Modified from Systems Design and Operational Effectiveness 625 Class Note—"Systems Design and Operational Effectiveness," Stevens Institute of Technology, 2007.

TABLE C-1 Hierarchy of Systems Components

Term	Definition
System of systems[a]	A configuration of systems in which component systems can be added or removed during use, each providing useful services in its own right, and each is managed for those services. Yet together they exhibit a synergistic, transcendent capability.
System[b]	An integrated set of elements, segments, and/or subsystems that accomplish a defined objective, such as an air transportation system.
Subsystem[b]	An integrated set of assemblies, components, and parts that performs a clearly separate function, involving similar technical skills, or a separate supplier. Examples are an aircraft onboard communications subsystem or an airport control tower as a subsystem of the air transportation system.
Assembly[b]	An integrated set of components and/or subassemblies that constitute a defined part of a subsystem, e.g., the pilot's radar display console on the fuel injection assembly of the aircraft propulsion subsystem.
Subassembly[b]	An integrated set of components and/or parts that comprise a well-defined portion of an assembly, e.g., a video display with its related integrated circuitry of a pilot's radio headset.
Component[b]	Composed of multiple parts; a clearly identified item, e.g., a cathode ray tube or the ear piece of the pilot's headset.
Part[b]	The lowest level of separately identified items, e.g., a bolt to hold a console in place.

[a]Air Force Scientific Advisory Board, 2005, *System-of-Systems Engineering for Air Force Capability Development*, SAB-TR-05-04, July. Available at http://stinet.dtic.mil/oai/oai?&verb=getRecord&metadataPrefix=html&identifier=ADA442612. Last accessed on April 2, 2007.

[b]International Council on Systems Engineering (INCOSE), 2004, *INCOSE Systems Engineering Handbook (Version 2A)*, Seattle, Wash.: INCOSE.

maintainability, supportability, global flexibility, scalability, interoperability, upgradability, and other special capabilities into the overall engineering effort.

Figure C-1 shows an important relationship between three parallel aspects of system development: the functional decomposition of a system shown in the center, supportability and logistics shown on the right, and cost shown on the left. As one follows each process flow, activities across each type of requirement are interdependent upon, and impact, one another. This figure demonstrates the importance of trade-off analysis in developing system requirements to balance performance, cost, and other specialties throughout the system life cycle.

SE goes well beyond traditional engineering concepts and tools. In the broadest sense, it encompasses systems thinking and other related systems disciplines

TABLE C-2 Standard Definitions of Systems Engineering (SE)

Source	SE Definition
International Council on Systems Engineering[a]	Systems engineering is an interdisciplinary approach and means to enable the realization of successful systems.
Military Standard on Engineering Management 499A[b]	The application of scientific and engineering efforts to: (1) transform an operational need into a description of system performance parameters and a system configuration through the use of an iterative process of definition, synthesis, analysis, design, test, and evaluation; (2) integrate related technical parameters and ensure compatibility of all related, functional, and program interfaces in a manner that optimizes the total system definition and design; and (3) integrate reliability, maintainability, safety, survivability, human, and other such factors into the total technical engineering effort to meet cost, schedule, and technical performance objectives.
Department of Defense[c]	Systems engineering is an interdisciplinary approach or a structured, disciplined, and documented technical effort to simultaneously design and develop systems products and processes to satisfy the needs of the customer. Systems engineering transforms needed operational capabilities into an integrated system design through concurrent consideration of all lifecycle needs.
NASA[d]	Systems engineering is a robust approach to the design, creation, and operation of systems.

[a]Benjamin Blanchard and Wolter Fabrycky, 2005, *Systems Engineering and Analysis* (4th Edition), Englewood Cliffs, N.J.: Prentice Hall.

[b]United States Air Force, 1974, *Military Standard—Engineering Management*, MIL-STD-499A, May 1, Washington, D.C.: Department of Defense.

[c]*Defense Acquisition Guidebook*, Chapter 4, Section 4.1.1. Available at https://akss.dau.mil/dag/DoD5000.asp?view=document&rf=GuideBook\IG_c4.1.1.asp. Last accessed on December 3, 2007.

[d]National Aeronautics and Space Administration, 1995, *Systems Engineering Handbook*, SP-610S, June. Available at http://snebulos.mit.edu/projects/reference/NASA-Generic/NASA-STD-8739-8.pdf. Last accessed August 30, 2007.

inherent to the execution of traditional engineering. It is not solely intended for those products described in the academic definition of a system but should also include subsystems, systems of systems, and enterprise-level problems. SE should be applied to all areas that affect the successful completion of a system, including financial management, management of technical risk, political support, and social context.

SE practices and approaches have historically been applied to everything from single systems to complex systems of systems. The SE community (e.g., the Institute of Electrical and Electronics Engineers, SoS conferences, the Inter-

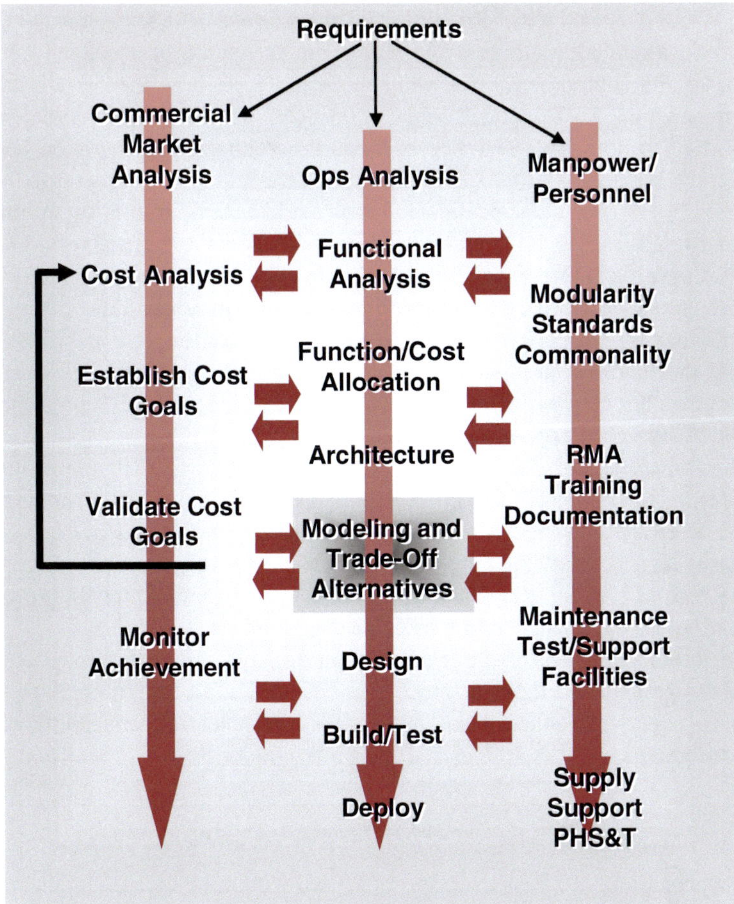

FIGURE C-1 Relationship among the traditional systems engineering functions (center column), cost (left column), and supportability and logistics (right column). SOURCE: Modified from Systems Design and Operational Effectiveness 625 Class Notes—"Systems Design and Operational Effectiveness," Stevens Institute of Technology, 2007.

national Council on Systems Engineering [INCOSE], and the System of Systems Center of Excellence) is paying increasing attention to issues of SoS, complex systems, and enterprise systems domains. In this report, the committee has loosely used the term SE to apply to tools, techniques, and processes for all of these domains.

Primary products of good SE are robust and efficient architectures. Architectures are multidimensional representations or combinations of "what, how, where, who, when, and why." Regardless of perspective, method, source data, and framework, an architecture description is a representation of a defined domain in terms of its component parts, what those parts do, how the parts relate to one another, and the rules and constraints under which the parts function. It is important to note the difference between an architecture description and an architecture implementation. An architecture description is a representation or "blueprint" of a current or postulated "real-world" configuration of resources, rules, and relationships. It generally contains "views" that are meaningful to each of the multiple disciplines involved in the implementation of the system. Once the blueprint enters the design, development, and acquisition process, the architecture description is then transformed into a real implementation of capabilities and assets in the field. An architecture framework provides guidance in describing architectures, but requires other tools in the tool set to move from representation to implementation of capabilities and assets.

The Department of Defense Architecture Framework (DDAF) shown in Figure C-2 is a three-dimensional representation of the multidimensional architecture space. For DDAF, the operational, technical, and system views are critical to developing an understanding of any system, and collectively these should capture the data derived from an analysis of the system. The "All Views" products provide information pertinent to the entire architecture.

An interdisciplinary effort (or team approach) is required throughout the system design and development process to ensure that all design objectives are met in an effective manner. This necessitates a complete understanding of the many different design disciplines and their interrelationships.

TOOLS AND METHODOLOGIES

A wide variety of SE methodologies is used within the defense industry, and an even larger collection of tools has sprung up to support SE processes in general. Information on commercial off-the-shelf (COTS) and government off-the-shelf (GOTS) tools of interest to systems engineers is available from the INCOSE Tools Database Working Group on its Web site.[3] The database categorizes the tools into four general areas:

- Requirements Management Tools Survey,
- Systems Architecture Tools Survey,
- Measurement Tools Survey, and
- General Tools Database.

[3]INCOSE Systems Architecture Tools Survey, available at http://www.paper-review.com/tools/sas/index.php. Last accessed on April 2, 2007.

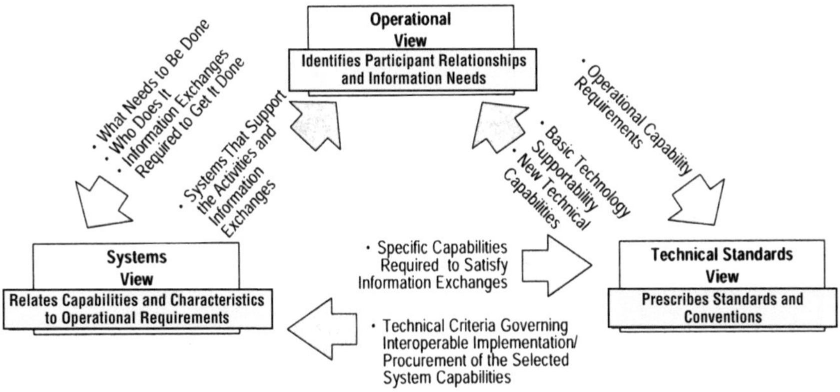

FIGURE C-2 Department of Defense Architecture Framework. SOURCE: William Wood, Mario Barbacci, Paul Clements, Steve Palmquist, Huei-Wan Ang, Loring Bernhardt, Fatma Dandashi, David Emery, Sarah Sheard, Lyn Uzzle, John Weiler, and Art Krummenoehl, 2003, *DOD Architecture Framework and Software Architecture Workshop Report*. Carnegie Mellon University: Software Engineering Institute. March. Available at http://www.sei.cmu.edu/publications/documents/03.reports/03tn006.html. Last accessed on April 2, 2007. Copyright 2003 by Carnegie Mellon University. Reprinted with special permission from the Software Engineering Institute.

Under the general heading of SE methodologies are the practices that promote and enable success. Some practices to be considered include the following:

- Well-documented processes, readily accessible to the users by being Web-based, searchable, and readily available online and in real time; can be tailored on a program-by-program basis prior to program baselining; and configuration can be managed under a synchronized, yet independent process on each program once the program is baselined;
- Templates for all engineering tasks organized by function and discipline, with examples of prior successes and links to available experts;
- Lists of available, compatible tool sets matched with the engineering tasks that they automate or support, with solved problems in a searchable archive;
- Lists of metrics appropriate to each task and program phase, with the mechanisms for automated collection, tools for analysis, examples of detected anomalies, and links to available experts with experience using the tools and in performing the analyses;
- Boilerplate work packages validated by prior usage and containing hints for tailoring and expediting, and parametric models of the effort required to

perform the work (cost, schedule, quality, inputs and outputs, required training/experience, and so on);

- A library of architectures and designs used on prior programs with links to the designers and implementers of each (both successful and unsuccessful);
- Anecdotes of problems actually discovered and fixes implemented, linked to relevant work package descriptions and to the individuals who performed the work; and
- A library of composable models of systems and subsystems (at multiple levels of detail) with links to designers; and for purchased items, to vendors or other available sources of supply.

MODELING AND SIMULATION

Modeling and simulation continue to be key elements of SE throughout the acquisition life cycle, especially early in programs. Modeling and simulation allow program managers to quickly develop concepts of operations (CONOPS) and analyses of alternatives as part of pre-Milestone A activities. Further along in the life cycle, modeling and simulation can be used for detailed design. Modeling and simulation can be used in a distributed collaborative environment that supports authoritative information exchange and rapid refinement of the design or concept, and over the system life cycle to respond to changing circumstances such as technological advances, changing threats, tactics, or doctrine. Much of the modeling and simulation activity during the pre-Milestone A period is the responsibility of the systems engineers and development planning experts in the government acquisition organization. In most cases, they will be supported by systems engineering and technical assistance (SETA) contractors and the federally funded research and development centers (FFRDCs) as they interact with users to fully understand what is needed, identify the existing systems that will coexist and interoperate with the new capabilities to be acquired, and build a modeling and simulation environment adequate to support the acquisition. During the early phases of defining the needed modeling and simulation environment, the SE team must also establish the metrics to be used for evaluating candidate concepts. The specifics of the modeling and simulation environment can then be filled in such a way that meaningful measures of merit can be extracted from the simulations and used to focus further rounds of simulation, critical experiments, and human-in-the-loop testing.

SYSTEMS ENGINEERING IN DOD ACQUISITION PROGRAMS AND PRE-MILESTONE A SYSTEMS ENGINEERING

With budgets becoming tighter, public scrutiny becoming stronger, the increasing focus being placed on advanced technology, and demands arising from the shift toward network-centric warfare, there has been a major emphasis placed on SE within DOD.[4] Policies such as the 5000 series[5] and the SoS guide [6] and the creation of the Systems and Software Engineering Office within the Office of the Under Secretary of Defense for Acquisition, Technology and Logistics (OUSD AT&L) point to an understanding of the contributions that SE can make to modern acquisition.

Figure C-3 shows the prescribed acquisition process prior to Milestone A as presented in DOD Instruction 5000.2.

Before a program can enter formally into the concept refinement phase, a concept decision milestone must be cleared. Typically this consists of an initial concept, an approved analysis of alternatives plan, and an established Milestone A date. After the concept decision, the concept development/refinement is used to refine the initial concept and to help reduce technical risk. The concept development phase is guided by the Initial Capabilities Document and AoA with continuous feedback to develop a technology development strategy. Modeling and simulation, optimization, and life cycle costing are all needed to conduct a meaningful analysis of alternatives. Systems engineering products such as the systems engineering plan (SEP)[7] have traditionally not been used prior to the Milestone A decision. In a policy memorandum dated February 20, 2004,[8] the ODUSD (AT&L) directed that the SEP become a requirement for each milestone review. The next version of the DOD 5000 series of acquisition documents will be updated to reflect this policy.

---

[4]Michael W. Wynne and Mark D. Schaeffer, 2005, "Revitalization of Systems Engineering in DoD," *Defense AT&L*: March-April, pp. 14-17.

[5]The 5000 series refers to DOD Directive 5000.1, "The Defense Acquisition System," and DOD Instruction 5000.2, "Operation of the Defense Acquisition System."

[6]Department of Defense (DOD), 2006, *System of Systems Systems Engineering Guide: Considerations for Systems Engineering in a System of Systems Environment, Version 9*, December 22. Available at http://www.acq.osd.mil/se/to%20be%20posted/SOSE%20Guide%20Dec%2022%20PDF.pdf. Last accessed June 26, 2007.

[7]DOD, 2006, *Systems Engineering Plan (SEP) Preparation Guide, Version 1.02*, February 10. Available at http://www.acq.osd.mil/se/publications/pig/sep_prepguide_v1_2.pdf. Last accessed June 12, 2007.

[8]Office of the Deputy Under Secretary of Defense (ODUSD) (AT&L), 2004, Policy memorandum entitled "Policy for Systems Engineering in DoD," Washington, D.C., February 20.

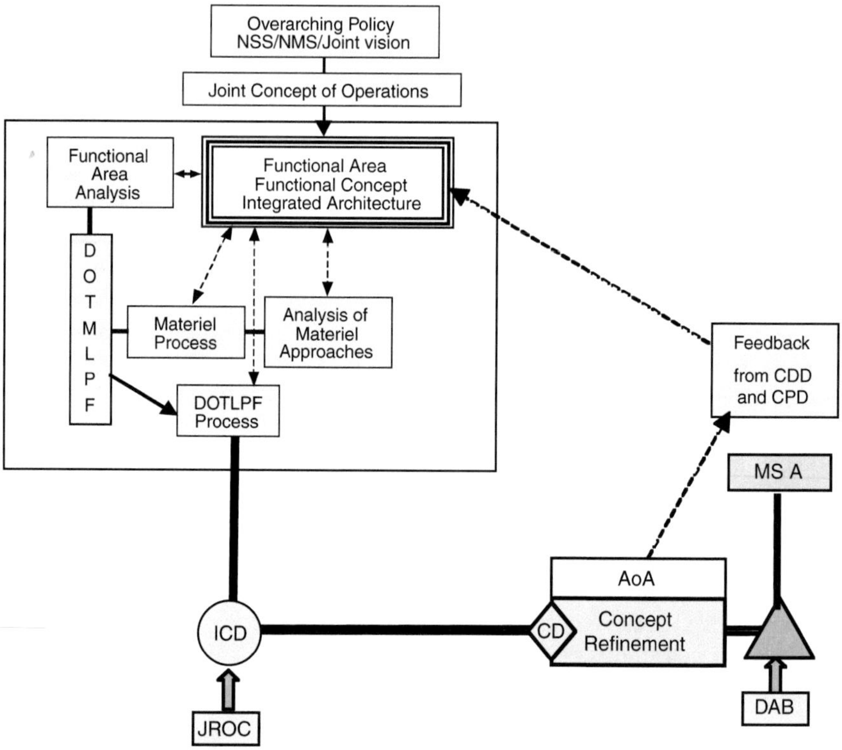

FIGURE C-3 Requirements and acquisition process prior to Milestone A. AoA, analysis of alternatives; CD, concept development; CDD, capabilities development document; CPD, Capabilities Production Document; DAB, Defense Acquisition Board; DOTLPF, doctrine, organization, training, leadership, personnel, and facilities; DOTMLPF, doctrine, organization, training, material, leadership and education, personnel, and facilities; ICD, Initial Capabilities Document; JROC, Joint Requirements Oversight Council; NSS, National Security Strategy; NMS, National Military Strategy; MS A, Milestone A. SOURCE: Department of Defense Instruction 5000.2, 2003, *Operation of the Defense Acquisition System.* May 12. Available at http://dod5000.dau.mil/DOCS/DoDI%205000.2-signed%20(May%2012,%202003).doc. Last accessed on April 2, 2007.